Memorix   Physiology

D1258383

Memorix

The *Memorix* series consists of easy to use pocket books in a number of different medical and surgical specialities. They contain a vast amount of practical information in very concise form through the extensive use of tables and charts, lists and hundreds of clear line diagrams, often two colours.

Memorix will give students, junior doctors and some of their senior colleagues a handy and comprehensive reference in their pockets.

Titles in the series include:

Obstetrics
Thomas Rabe

Gynecology
Thomas Rabe

Emergency Medicine
Sönke Müller

Neurology
Peter Berlit

Clinical Medicine
Conrad Droste and Martin von Planta

Surgery
Jürgen Hussmann and Robert Russell

Physiology
Robert F. Schmidt, W.D. Willis and L. Reuss

Pediatrics
Dieter Harms and Jochem Scharf

Memorix

Physiology

Robert F Schmidt
William D Willis
Luis Reuss

CHAPMAN & HALL MEDICAL

London · Weinheim · New York · Tokyo · Madras

Published by Chapman & Hall, 2–6 Boundary Row, London SE1 8HN, UK

Chapman & Hall, 2–6 Boundary Row, London SE1 8HN, UK

Chapman & Hall GmbH, Pappelallee 3, 69469 Weinheim, Germany

Chapman & Hall USA, 115 Fifth Avenue, New York, NY 10003, USA

Chapman & Hall Japan, ITP-Japan, Kyowa Building, 3F, 2-2-1 Hirakawacho, Chiyodaku, Tokyo 102, Japan

Chapman & Hall India, R. Seshadri, 32 Second Main Road, CIT East, Madras 600 035, India

English language edition 1997

© 1997 Chapman & Hall

Modified from Physiologie kompakt;
2. Auflage, © 1995, Gustav Fischer Verlag

Typeset in Times by Best-set Typesetter Ltd, Hong Kong
Printed and bound in Hong Kong

ISBN 0 412 71440 X

A catalogue record for this book is available from the British Library

Library of Congress Catalog Card Number: 97-66861

Contents

CONTENTS

CONTENTS

CONTENTS

Preface

This English edition of *Memorix Physiology* is intended to provide for English-speaking students of physiology the same advantages as did the original German edition for its German-speaking readership. The book gives an overview of the essential concepts of medical physiology in an easily understood format and with ample illustrations.

The English translation was prepared by Dr Marguerite Biederman-Thorson. Luis Reuss and William D Willis adjusted the translation in keeping with terminology used by physiologists in the United States, and they joined Robert F. Schmidt in preparing what amounts to a new edition of the book. Besides a number of additions to and changes in the text, the sequence of many of the chapters has been altered.

Several members of the University of Texas Medical Branch faculty assisted Luis Reuss by reviewing material in their specialty areas. These individuals include Drs David J Bessman, Malcolm S Brodwick, Gwendolyn V Childs, Bryan O Holland, Giuseppe Sant' Ambrogio, Donald W Stubbs and Steven A Weinman. The authors thank them for their special effort.

Luis Reuss and William D. Willis, Galveston,
Texas
Robert F. Schmidt, Würzburg, Germany
May 1997

Short contents

Physiological units of measurement

Prefixes and symbols for frequently used powers of ten

Factor	Prefix	Symbol	Factor	Prefix	Symbol
10^{-1}	deci	d	10	deca	da
10^{-2}	centi	c	10^2	hecto	h
10^{-3}	milli	m	10^3	kilo	k
10^{-6}	micro	µ	10^6	mega	M
10^{-9}	nano	n	10^9	giga	G
10^{-12}	pico	p	10^{12}	tera	T
10^{-15}	femto	f	10^{15}	peta	P

Names and symbols of the International System base units

Quantity	Name of unit	Symbol
1. Length	meter	m
2. Mass	kilogram	kg
3. Time	second	s
4. Electric current	ampere	A
5. Absolute temperature	kelvin	K
6. Luminous intensity	candle	cd
7. Amount of a substance	mole	mol

The units of all of the other quantities in the International System (see below) are derived from the above seven base units. Commonly used powers of ten are designated by adding to these units standard prefixes and symbols as shown above. But other base units are also in use (see below). Some conventional units (e.g. HP, kcal, mmHg) tenaciously persist along with the International System units (W, J, Pa).

Names and symbols of some derived International System units

Quantity	Name	Symbol	Definition
Frequency	hertz	Hz	s^{-1}
Force	newton	N	$m \cdot kg \cdot s^{-2}$
Pressure	pascal	Pa	$m^{-1} \cdot kg \cdot s^{-2}$ $(N \cdot m^{-2})$
Energy	joule	J	$m^2 \cdot kg \cdot s^{-2}$ $(N \cdot m)$
Power	watt	W	$m^2 \cdot kg \cdot s^{-3}$ $(J \cdot s^{-1})$
Electric potential difference	volt	V	$m^2 \cdot kg \cdot s^{-3} \cdot A^{-1}$ $(W \cdot A^{-1})$
Electric resistance	ohm	Ω	$m^2 \cdot kg \cdot s^{-3} \cdot A^{-2}$ $(V \cdot A^{-1})$

Commonly used base units not belonging to the International System

Quantity	Name	Symbol	Definition
Mass	gram	g	$1\ g = 10^{-3}\ kg$
Volume	liter	l	$1\ l = 1\ dm^3$
Time	minute	min	$1\ min = 60\ s$
Time	hour	h	$1\ h = 3600\ s$
Time	day	d	$1\ d = 86\,400\ s$ $(8.64 \cdot 10^4\ s)$
Temperature	degrees Celsius	°C	$0°C = 273.15\ K$

Important conversions between International System and conventional units

Quantity	Conversion equations	
Force	$1\,dyn = 10^{-5}\,N$ $1\,kp = 9.81\,N$	$1\,N = 10^5\,dyn$ $1\,N = 0.102\,kp$
Pressure	$1\,cmH_2O = 9.81\,Pa$ $1\,mmHg = 133\,Pa$ $1\,atm = 101\,kPa$ $1\,bar = 100\,kPa$	$1\,Pa = 0.0102\,cmH_2O$ $1\,Pa = 0.0075\,mmHg$ $1\,kPa = 0.0099\,atm$ $1\,kPa = 0.01\,bar$
Energy (work) (amount of heat)	$1\,erg = 10^{-7}\,J$ $1\,m\,kp = 9.81\,J$ $1\,cal = 4.19\,J$	$1\,J = 10^7\,erg$ $1\,J = 0.102\,m\,kp$ $1\,J = 0.239\,cal$
Power (heat flow) (energy metabolism)	$1\,m\,kp/s = 9.81\,W$ $1\,HP = 736\,W$ $1\,kcal/h = 1.16\,W$ $1\,kJ/d = 0.0116\,W$ $1\,kcal/d = 0.0485\,W$	$1\,W = 0.102\,m\,kp/s$ $1\,W = 0.00136\,HP$ $1\,W = 0.860\,kcal/h$ $1\,W = 86.4\,kJ/d$ $1\,W = 20.6\,kcal/d$
Viscosity	$1\,poise = 0.1\,Pa.s$	$1\,Pa\,s = 10\,poise$

General cell physiology

Scheme of the structure of a human cell

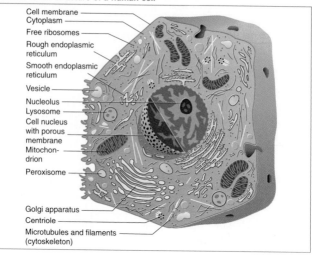

Cell membrane
Cytoplasm
Free ribosomes
Rough endoplasmic reticulum
Smooth endoplasmic reticulum
Vesicle
Nucleolus
Lysosome
Cell nucleus with porous membrane
Mitochondrion
Peroxisome
Golgi apparatus
Centriole
Microtubules and filaments (cytoskeleton)

Number of cells in the human body

Cell type	Number	Comment
Total	$75 \cdot 10^{12}$	i.e. 75 trillion
Erythrocytes	$25 \cdot 10^{12}$	the most abundant cell type
Neurons	$25 \cdot 10^{9}$	i.e. 25 billion

Most important cell components (see Fig.)

Component	Description/comments
Cell membrane	Synonym for plasma membrane (q.v.) or plasmalemma
Centriole	Short cylinder of microtubules; involved in cell division
Chromatin	DNA protein (chromosomes) of the cell nucleus (nuclear 'skeleton') altered by fixation and staining
Cytoplasm	The ensemble of cytosol, cytoskeleton and cytoplasmic organelles (e.g. endoplasmic reticulum, Golgi apparatus, ribosomes)
Cytoskeleton	Network of protein filaments in the cytoplasm, mainly actin filaments (microfilaments), microtubules and intermediate filaments
Cytosol	Gelatinous (20% protein) cell fluid with thousands of enzymes; involved in intermediary metabolism
Endoplasmic reticulum	Elementary membranes arranged in a labyrinth of passages, clefts and tubes ('rough' ER if ribosomes present); involved in metabolism, e.g. protein synthesis
Golgi apparatus	Stack of 5–30 cisternae formed by smooth membranes; involved in intracellular protein and lipid transport
Lysosome	Cell organelle, product of the Golgi apparatus; digestion of material originating both within and outside the cell
Microtubule	Part of the cytoskeleton made of protein with tubular structure; participates in cell division, axonal transport, movement of cilia and flagella
Mitochondrion	Membrane-bound organelle abundant in cells with high energy consumption; site of synthesis of ATP; contains some DNA
Nucleus	Contains the chromosomes (which carry genetic hereditary information encoded by DNA molecules). A membrane encloses the kartiplasm (nucleoplasm)
Peroxisome	Membrane-bound organelle, notably abundant in liver and kidney cells; involved in metabolism
Plasma membrane	Consists mainly of phospholipid (e.g. phosphatidyl choline) with hydrophilic head groups and hydrophobic fatty acid chains; these form a 4–5 nm thick bilayer (hydrophilic head groups toward the water), in which proteins are embedded as the chief functional elements (channels, carriers, pumps, receptors)
Ribosome	Organelle, part of the rough ER; translation system in protein synthesis
Vesicle	Membrane-bound organelle; involved in endocytosis, exocytosis, transcytosis

Main routes for exchange of substances between the cell and its surroundings

Passive transport through the plasma membrane	
Diffusion 1. through lipid membrane	Particle movement along concentration gradients Direct passage through lipid membrane possible only for substances that are both water- and lipid-soluble; e.g. fatty acids, O_2, alcohol; follows Fick's law of diffusion
2. through membrane channels	Movement through the water-filled channels formed by integral pores membrane proteins; depends on concentration gradient, membrane voltage, particle charge in the case of ions and channel selectivity
Facilitated transport	Binding to transport protein on the membrane causes change in conformation, releasing molecule on other side of membrane; net movement along electrochemical gradient
Carrier (transporter)	Transport protein that facilitates diffusion across the plasma membrane (uniporter); e.g. glucose, amino acids
Cotransporter	Transport protein that facilitates transport of two or more substrates in the same direction; e.g. Na^+–glucose, Na^+–K^+–$2Cl^-$
Exchanger (anti-porter)	Transport protein that allows exchanges of substances across the plasma membrane; e.g. Na^+–Ca^{2+}, Na^+–H^+, Cl^-–HCO_3^-
Solvent drag	Water 'drags along' dissolved substances as it passes barriers

Active transport ('pumping') through the plasma membrane	
Electroneutral pump	Primary active transport of particles against the electrochemical gradient; energy is expended (ATP hydrolysis); no net transport of electric charge
Electrogenic pump	Primary active transport in which there is a shift in net electric charge; e.g. Na^+, K^{+}-ATPase ($3Na^+ : 2K^+$), the most important pump in the animal cell membrane
Secondary active transport	An electrochemical gradient created by pump activity is exploited so that carriers (transporters) can move substances in the same (cotransport or symport) or the opposite direction (countertransport or antiport). See above
Exocytosis	Fusion of intracellular vesicles with the plasma membrane, resulting in release of their content (e.g. hormone) into the external medium
Endocytosis	Transport of solid (phagocytosis) or dissolved (pinocytosis) substances into cell by invagination and pinching off of plasma membrane; receptor-mediated endocytosis involves clathrin-coated pits, which give rise to coated vesicles

Main routes for exchange of substances inside the cell

Exchanges at intracellular membranes	
Diffusion	Processes identical to those at the plasma membrane (see above); total surface area of the organelle membranes is at least ten times as large as that of the plasma membrane
Active transport	Processes identical to those at the plasma membrane (see above); important examples: ATP synthesis at mitochondrial membrane; Ca^{2+} pump in the sarcoplasmic reticulum

Transport processes in the cytoplasm	
Diffusion	In the cytosol, eliminates concentration differences of dissolved particles; slower than in pure water because of viscosity due to high protein concentration in cytosol
Transport in vesicles	Involved in endocytosis (see above), exocytosis (see above) and vesicle transcytosis (transport within the cell)
Axonal transport 1. Fast	Transport of substances in nerve fibers (axons) Vesicles and mitochondria are moved along microtubules by attaching to a transport protein. Process depends on energy, Ca^{2+} and a protein motor. Anterograde transport is toward the periphery, involves kinesin; rate up to 400 mm/day. Retrograde transport is toward soma, involves dynein and is up to 200 mm/day
2. Slow	Bulk transport of substances, especially proteins, dissolved in cytosol. About 1 mm/day (rate of regeneration of axons)

4

Information exchange (signal transmission, communication) between cells

Information exchange between cells is characteristic of the nervous system and the endocrine system. The final step in transmission of information in both systems is usually or always chemical and involves release of neurotransmitters in the nervous system and hormones in the endocrine system. These first messengers bind to the plasma membrane or to intracellular receptors to alter the function of cells receiving the message. They may open membrane channels or they may have a complex effect through activation of second messenger systems. When the action of a released messenger is restricted to cells in the immediate vicinity of the release site (within about 1 cm), the process is called paracrine signal transmission.

The main low molecular weight ('classical') neurotransmitters

Name	Location(s) of cells	Synaptic release site(s)
Acetylcholine	Motoneurons Preganglionic autonomic neurons Postganglionic parasympathetic and some postganglionic sympathetic neurons CNS	Neuromuscular junction Autonomic ganglia Smooth muscle, glands, cardiac muscle Brain, spinal cord
Norepinephrine	Postganglionic sympathetic neurons CNS (locus coeruleus, subcoeruleus)	Smooth muscle, glands, cardiac muscle Brain, spinal cord
Epinephrine	Adrenal medulla CNS (brainstem)	Hormonal action on smooth muscle, glands, cardiac muscle, fat cells Brain, spinal cord
Dopamine	CNS (substantia nigra, hypothalamus)	Brain (striatum, median eminence, limbic system)
Serotonin	CNS (raphe nuclei)	Brain, spinal cord
Histamine	Mast cells CNS	Local hormonal action on smooth muscle, glands Brain

'Classical' transmitters are so called because they have long been recognized as such. In recent years, several amino acids have also been elevated to the status of 'classical' transmitters, and a number of peptides have also been discovered to play a role in synaptic transmission (the same peptides are often also hormones). 'Classical' transmitters are frequently co-localized in nerve endings with peptides and are presumed to be co-released and to exert cooperative actions.

Amino acids (now 'classical' neurotransmitters)

Name	Location(s) of cells	Synaptic release site(s)
Glutamate	Brain, spinal cord, dorsal root and cranial nerve ganglia	Brain, spinal cord
Aspartate	Brain, spinal cord	Brain, spinal cord
γ-Aminobutryric acid (GABA)	Brain, spinal cord	Brain, spinal cord
Glycine	Spinal cord	Spinal cord

Peptides are found co-localized with 'classical' neurotransmitters in synapses in the PNS, brain and spinal cord. Many of the same peptides are hormones found in the hypothalamus, pituitary gland, gastrointestinal tract, adrenal medulla and other sites.

Type

Hypothalamic releasing or inhibiting hormones
Corticotropin-releasing hormone (CRH); gonadotropin-releasing hormone (GnRH); growth hormone-releasing hormone (GHRH); growth hormone-inhibiting hormone (GHIH, somatostatin); melanocyte-stimulating hormone-releasing hormone (MSH-RH); melanocyte-stimulating-inhibiting hormone (MSH-IH); prolactin-releasing hormone (PRL-RH); prolactin-releasing-inhibiting hormone (PRL-IH); thyrotropin-releasing hormone (TRH)

Hormones of the neurohypophysis
Antidiuretic hormone (ADH; or arginine vasopressin, AVP); oxytocin

Hormones of the adenohypophysis
Adrenocorticotropic hormone (ACTH); β-endorphin; follicle-stimulating hormone (FSH); growth hormone (GH; or somatotropic hormone, STH); interstitial cell-stimulating hormone (ICSH); luteinizing hormone (LH); melanocyte-stimulating hormone (α-MSH); prolactin; thyrotropic hormone (TSH)

Gastrointestinal peptides
Bombesin; calcitonin gene-related peptide (CGRP); cholecystokinin (CCK); enkephalin (ENK); glucagon; insulin; motilin; neuropeptide Y (NPY); neurotensin (NT); secretin; somatostatin; substance P (SP); vasoactive intestinal polypeptide (VIP)

Others
Angiotensin II; bradykinin; dynorphin

Lock-and-key concept of the interaction between a messenger substance (transmitter, hormone) and a receptor and the actions of agonists and antagonists

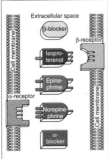

The binding of the 'key' (messenger substance) to the 'lock' (receptor) is the information-transmitting process in humoral communication. Each particular system, however, employs somewhat different 'keys' and 'locks'.

Agonists of a messenger substance are chemical agents that bind to the same receptor and exert the same effect (e.g. postsynaptic excitatory or inhibitory).

Antagonists are chemical agents that also bind to a receptor, but they block the action of the messenger substance or its agonists.

Agonists and antagonists may not react with all receptors for a particular messenger substance; the differences in action of these agents are used for pharmacological identification of various receptor types and are exploited therapeutically.

A messenger substance may bind to a receptor associated with a ligand-gated ion channel, to a receptor associated with a G-protein or to a tyrosine kinase.

Receptors associated with ligand-gated ion channels

Receptor type (transmitter)	Properties; locations
Nicotinic (acetylcholine)	Cation channel; motor end-plates on skeletal muscle, autonomic ganglia; also in CNS
5-HT$_3$ (serotonin)	Cation channel; neurons in PNS and CNS
Ionotropic glutamate receptors AMPA (glutamate) Kainate (glutamate) NMDA (glutamate)	Cation channel; CNS neurons Cation channel; CNS neurons Voltage-dependent cation channel; allows Ca^{2+} influx; blocked by Mg^{2+}; glycine is co-agonist; in CNS; involved in neuroplasticity
GABA$_A$ (GABA)	Cl^- channel; major inhibitory receptor in CNS
Glycine (glycine)	Cl^- channel; important inhibitory receptor in CNS

Other receptors involve GTP-binding proteins known as G-proteins. G-proteins are activated when a messenger substance binds to its receptor. The activated G-protein in turn can alter the activity of ion channels or enzymes. Activated G-proteins have an increased affinity for GTP, whereas inactivated G-proteins have a decreased affinity for GDP. GTP binds to activated G-protein, which then hydrolyzes the GTP to GDP, eventually returning the G-protein to its inactive state. G-proteins can be stimulatory (G_s) or inhibitory (G_i) in their effects.

Receptors associated with G-proteins

Receptor type (transmitter)	Properties; locations
Muscarinic (acetylcholine)	Act on K^+, Ca^{2+} and Cl^- channels and the enzymes adenylyl cyclase, phospholipase C, D and A_2, and guanylyl cyclase; found on cells of heart, smooth muscle, glands, and neurons in autonomic ganglia, CNS
Adrenergic (epinephrine, norepinephrine)	
α_1-adrenergic	Stimulate phospholipase C to increase level of inositol triphosphate, which releases Ca^{2+} intracellularly; may also open Ca^{2+} channels; found in smooth muscle, heart, liver, kidney, CNS
α_2-adrenergic	Act on K^+ channels; also inhibit adenylyl cyclase; can be pre- or postsynaptic in CNS; also found on platelets and in lung
β_1-adrenergic	Stimulate adenylyl cyclase; found in heart, kidney, CNS
β_2-adrenergic	Stimulate adenylyl cyclase; found in lung, blood vessels, kidney, uterus
Dopaminergic (dopamine)	
D_1	Stimulate adenylyl cyclase; found in CNS, especially in the basal ganglia and basal forebrain
D_2	Inhibit adenylyl cyclase; found in CNS, especially in the basal ganglia, mesolimbic system and pituitary
Serotonergic (serotonin)	
$5-HT_{1A}$	Inhibit adenylyl cyclase; can activate phospholipase C to increase IP_3 level; may be pre- or postsynaptic; found in CNS, especially limbic system, raphe nuclei
$5-HT_2$	Increase IP_3 levels; activate phospholipase A_2; CNS, including cerebral cortex, basal ganglia
Metabotropic (glutamate)	
mGluR1	Stimulate phospholipase C, adenylyl cyclase, phospholipase A_2; found in CNS, including cerebellum, hippocampus, brainstem
mGluR2	Inhibit adenylyl cyclase; found in CNS, including hippocampus

Other first messengers using G-protein-linked receptors: histamine; adenosine; cytokines; opioids; tachykinins

Second messenger systems serve to link the first messengers (transmitters and hormones) to intracellular responses (actions on ion channels, enzymes or gene expression). Examples of second messengers: Ca^{2+}, cAMP, cGMP, IP_3, DAG, nitric oxide (NO) and perhaps other gases.

Schematic organization of second messenger cascade with examples for cAMP and for IP_3 (adapted from Kandel *et al.*, 1991)

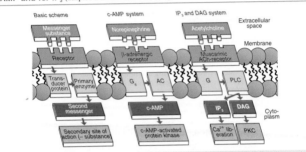

The diagram at the left shows the basic scheme for second messenger cascades. In step 1, a first messenger activates a receptor. This leads to step 2, the activation of a transducer protein, such as a G-protein. In step 3, the transducer protein activates a primary enzyme. This enzyme in step 4 produces the second messenger, which in step 5 either activates another enzyme or acts on a regulatory protein. The middle diagram shows the cAMP cascade. Norepinephrine acting at a β-adrenergic receptor causes a G_s-protein to stimulate adenylyl cyclase, which converts ATP to cAMP. cAMP then activates a protein kinase, which catalyzes the phosphorylation of proteins (e.g. ion channels, other enzymes). The diagram at the right shows the activation of a G-protein by acetylcholine acting at a muscarinic receptor. The G-protein activates phospholipase C (PLC), which converts phosphatidylinositol bisphosphate (PIP_2) into IP_3 and diacylglycerol (DAG). IP_3 releases Ca^{2+} from the endoplasmic reticulum, and DAG activates protein kinase C (PKC). PKC phosphorylates proteins. Not shown in either example is the terminate phosphorylation by phosphatases.

Some receptors are associated with tyrosine kinases, rather than with G-proteins. Activation of tyrosine kinases can cause Ca^{2+} influx, increase exchange or transport across the membrane, or activate second messenger cascades.

Examples of first messengers and receptors linked to tyrosine kinases

Insulin; insulin-related growth factors; epidermal growth factor; platelet-derived growth factor; fibroblast growth factor

Survey of the basic phenomena

Brief definitions of the basic concepts of cell excitation (types of potentials)

Term	Definition/description/comments
Membrane potential difference	General term for all electrical potential differences across cell membranes between the cell interior and the extracellular space (four basic types given below); intracellular, direct measurement with microelectrodes, diverse extracellular methods (e.g. ENG, ECG, EEG, EMG, depending on application)
Resting potential in Fig. ①	Membrane potential of excitable cells (nerve and muscle cells) in resting state; cell interior always negative with respect to extracellular space; magnitude −55 to −100 mV, depending on cell type. Inexcitable cells (e.g. glial cells) also have a resting potential
Action potential in Fig. ②	Brief, stereotyped ('all-or-none') change of membrane potential in the positive direction during excitation of cell; amplitude about 100 mV; duration: nerve and skeletal muscle cells ca. 1–2 ms, cardiac muscle cells ca. 350 ms (see p. 158)
Electrotonic potential in Fig. ③	(Synonym electrotonus) Positive (depolarizing) or negative (hyperpolarizing) departure from resting potential due to flow of current across the membrane. The depolarizing current acts as a stimulus; i.e. it triggers an action potential when a threshold is reached
Synaptic potential in Fig. ④	Depolarizing (excitatory) or hyperpolarizing (inhibitory) departure from resting potential due to activation of excitatory of inhibitory synapses (details on p. 17)

Intracellular measurement of membrane potential

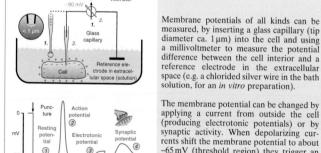

Membrane potentials of all kinds can be measured, by inserting a glass capillary (tip diameter ca. 1 μm) into the cell and using a millivoltmeter to measure the potential difference between the cell interior and a reference electrode in the extracellular space (e.g. a chlorided silver wire in the bath solution, for an *in vitro* preparation).

The membrane potential can be changed by applying a current from outside the cell (producing electrotonic potentials) or by synaptic activity. When depolarizing currents shift the membrane potential to about −65 mV (threshold region) they trigger an action potential

Molecular basis of the resting potential

Intracellular ion concentrations in a muscle cell of a warm-blooded animal, in comparison to the ion concentrations in the extracellular space, to illustrate the inequalities between these two spaces. A⁻ stands for 'large intracellular anions'

Intracellular space		$C_i : C_e$	Extracellular space	
Na^+	12 mmol/l	1 : 12.1	Na^+	145 mmol/l
K^+	155 mmol/l	39 : 1	K^+	4 mmol/l
Ca^{2+}	10^{-8}–10^{-7} mol/l		Ca^{2+}	2 mmol/l
Cl^-	4 mmol/l	1 : 30	Cl^-	120 mmol/l
HCO_3^-	8 mmol/l	1 : 3.4	HCO_3^-	27 mmol/l
A^-	155 mmol/l		Other cations	5 mmol/l
Resting potential −90 mV				

Conspicuous features of the cell interior are the high K^+ and the low Na^+ concentrations; in the extracellular space this relationship is reversed. Furthermore, outside the cell the Cl^- concentration is considerably higher than inside; in turn, the concentration of large anions is high intracellularly, but not outside. There are hardly any free Ca^{2+} ions in the cytoplasm of the cell.

The unequal distribution of Na^+ and K^+ ions across the cell membrane has 2 causes: pump activity and selective permeability

Term	Description/comments
Na^+,K^+-ATPase pump	Na^+,K^+-ATPase exchange pumps are densely packed within the cell membrane (see also p. 255, Fig. p. 258). These establish the very low Na^+ concentration and high K^+ concentration inside the cell and maintain it for the life of the cell. Each pump cycle moves 3 Na^+ out of the cell and 2 K^+ into it. The activity of the pumps is regulated by the intracellular Na^+ concentration; at ca. 10 mmol/l the pump becomes inactive
Selective permeability	The plasma membrane contains many channels (pores) that are permeable only to K^+ ions and are usually open at rest, so that at rest the K^+ permeability is very high. The membrane also contains many channels permeable only to Na^+ ions, but at rest these are almost all closed: the resting Na^+ permeability is very low. The membrane is impermeable to the large anions

The selective permeability of the membrane channels derives from their configuration and the electric charges in the wall of the pore

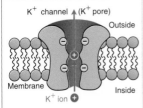

The K^+ channel allows K^+ ions to pass through but prevents passage of the positively charged and similarly sized Na^+ and Ca^{2+} ions: the 4 negative charges in the pore wall block the passage of Cl^- and other anions; on its way through the channel, K^+ binds to these charges. Evidently only K^+ can bind in this way, which is the basis of selectivity (the same principle applies to other pores, such as those for Na^+).

The unequal distribution of ions across the cell membrane and the high K⁺ permeability at rest produce the resting potential, which is basically a K⁺ diffusion potential

If the cell membrane were only permeable to K^+, then following the concentration gradient, the K^+ ions would tend to diffuse out of the cell, removing positive charge. Therefore the interior of the cell becomes negatively charged relative to the extracellular space, the resting potential. Its final value is reached when the electrical gradient associated with the negative charge (which 'pulls' the K^+ ions back into the cell) exactly balances the diffusion pressure along the concentration gradient. This (resting) potential of about -90 mV is thus the K^+ equilibrium potential.

In addition to the concentration gradient and the charge number z of the ion (negative for anions), the gas constant R, the absolute temperature T and the Faraday constant F determine the equilibrium potential of an ion:

$$E_i = R \cdot T/z \cdot F \left[\ln \left(C_0/C_i \right) \right] \qquad (1)$$

This is the general form of the Nernst equation. For body temperature (T = 310 K) it gives the following K^+ equilibrium potential:

$$E_K = -61 \, mV \cdot \log \left(\left[K^- \right]_{in} \Big/ \left[K^- \right]_{out} \right) \qquad (2)$$

Note transformation from natural to decimal logarithms.
For $[K^+]_{in}/[K^+]_{out} = 39$ (see Table) this becomes $E = -61 \, mV \cdot 1.59 = -97 \, mV$, a value demonstrating that the resting potential closely approximates but is not identical to the K^+ equilibrium potential. Therefore, even at the resting potential there is a continuous net outflow of K^+ ions from the cell.

The contributions of the Na⁺,K⁺-ATPase pump, of the other ions and of glial cells to the resting potential

Term	Action on resting potential/comments
Na⁺,K⁺-ATPase pump	The Na⁺,K⁺ pump translocates net positive charge out of the cell (see p. 255). This pump current contributes about -10 mV to the resting potential: if the pump is blocked by digitalis glycosides, the resting potential is rapidly reduced by this amount
Na⁺ ions	Because of the large concentration gradient (see Table p. 11) and the equilibrium potential of $E_{Na} = +60$ mV (according to Equation 1), despite the low Na⁺ permeability there is a continuous slight Na⁺ influx, which shifts the resting potential in the positive direction. The continuous Na⁺ influx and the continuous K⁺ efflux are opposite to fluxes mediated by the Na⁺,K⁺ pump
Cl⁻ ions	The Cl⁻ permeability of most cells is low, but that of muscle cells is high. The equilibrium potential for Cl⁻ ions is about the same as the resting potential, so that at rest there are no net Cl⁻ currents
Ca²⁺ ions	By means of Na⁺,Ca²⁺ antiport (see Fig. p. 258) the free intracellular Ca²⁺ concentration is kept at the low value shown in the Table on p. 11. The resting potential is not affected by this transport. However, Ca²⁺ currents contribute to the action potential in many cells, e.g. in the heart (see p. 158)
Glial cells	These have an almost pure potassium equilibrium potential. They are in electrical communication with one another by way of gap junctions. Local increases in K⁺ concentration (e.g. when the tissue is highly excited) can be buffered by the uptake of K⁺ into these cells; hence they act to stabilize the local milieu

Molecular basis of the action potential

Triggering, components and time course of the action potential (AP)

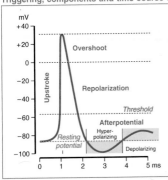

Action potentials (amplitude ca. 100 mV) are generated whenever the membranes of excitable cells are depolarized by about 20 mV from resting potential.

The depolarizing membrane current is called the stimulus and the potential level at which the action potential is triggered is called the threshold.

The various components and the time course of a neuronal action potential with an overall duration of 1–2 ms (similar to muscle action potential) are shown in the diagram (for time course of a cardiac action potential see p. 160). During an AP a cell is excited.

Ion currents during the action potential, mechanism of repolarization

The depolarization of the resting potential to the threshold initiates an abrupt opening of many Na^+ channels (gating). The resulting Na^+ influx drives the depolarization further (opening of additional Na^+ channels, 'all-or-none' character of the AP) and makes the interior of the cell positive. The return to the resting potential, called repolarization, occurs because (1) the Na^+ channels remain open for only about 1 ms and then close (inactivation); (2) the number of open K^+ channels also rises briefly above the resting value (so that K^+ ions diffuse out of the cell in larger amounts).

Behavior of single Na^+ channels, refractoriness

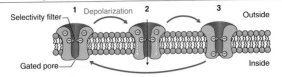

The Na^+ channel protein contains a selectivity filter that allows Na^+ (but not K^+) to permeate, as well as a voltage-gated pore that defines the opening state of the channel. The gated pore has three main states (the transitions are caused by changes in conformation of the channel protein): (1) closed–activatable, (2) open and (3) closed-inactivated. The transition from 1 to 2 occurs at random with an average probability that is low at rest but is increased by depolarization (potential dependence of the channels). State 3 necessarily follows 2 and then leads to 1 (time dependence of the channels). At rest with a high resting potential nearly all the channels are in state 1 ('Na fully activatable'); as the resting potential decreases, progressively more channels are converted to state 3, and finally the cell becomes inexcitable (at ca. -60 mV, e.g. in O_2 deficiency). As a consequence of inactivation of the Na system (state 3), the cell is inexcitable for a few ms after an AP; an absolutely refractory phase of complete inexcitability is followed after 1–2 ms by a relative phase (cell excitable but with elevated threshold and smaller AP).

Supplementary facts and definitions related to the action potential

Aspect	Description/comments
Structure and number of Na channels	Chemically the channel is a glycoprotein, MW ca. 300 000, diameter ca. 8 nm, channel (pore) width ca. 0.5 nm; number: $1–50/\mu m^2$, depending on type of membrane. Anions cannot pass through because of negative fixed charges at channel entrance. Opening by displacement of fixed membrane charges (gating current)
Pharmacology of the Na channel	Tetrodotoxin (TTX) blocks pore entrance, local anesthetics (e.g. lidocaine) become firmly lodged in the channel when it is open, entering from the cytoplasmic side and causing a use-dependent block. Inactivation can be prevented by toxins and drugs (e.g. pronase, iodate) that act from inside the cell
Analysis of the Na channel behavior	The overall ion flux can be studied with voltage clamp methods, and the behavior of individual channels with the patch clamp method. Under depolarization the channel molecules suddenly change their conformation and open for an average of 0.7–1.0 ms (state 2, see p. 13). During this time a current of about 1.6 pA flows; i.e. about 10 000 Na^+ ions pass through the channel. Closure of the channel causes brief inactivation (state 3)
Behavior of the K channel	Resembles the Na channel in many respects but is not inactivated during depolarization; instead it oscillates between states 1 and 2 for the entire duration of depolarization. However, there are very different types of K channels in different tissues. Therefore repolarization and afterpotentials are extremely variable
Conductances g_{Na} and g_K	Quantitative measure of the 'overall permeability' of a membrane to these two ions; hence applies to the total number of open channels at a given moment; by analogy with Ohm's law, is defined as the ratio of flowing current to driving voltage. Time course during the action potential shown in the figure on p. 15; at the resting potential g_K is 10–25 times greater than g_{Na}

Special features of Ca^{2+} ions and Ca channels

Aspect	Description/comments
Ca^{2+} ion concentration and excitability	Ca^{2+} ions 'stabilize' the resting potential by raising the threshold for activation of the fast Na system; conversely, reduction of the extracellular concentration of Ca^{2+} causes overexcitability, the extreme form of which is tetany (muscle cramps)
Behavior of the Ca channel	In nerve and skeletal muscle cells the Ca channel behaves like the Na channel, but since $g_{Ca} \ll g_{Na}$, it is irrelevant to normal excitation processes. At smooth and cardiac muscle cells, however, Ca^{2+} influx often exceeds Na^+ influx (Ca^{2+} action potential). The inflowing Ca^{2+} ions can serve intracellular control functions (e.g. activation of second messenger systems, excitation–contraction coupling)
Ca channel modulation	The open probability of the Ca channel can be increased, e.g. in the heart, by epinephrine and norepinephrine; the result is a larger Ca^{2+} current and hence improved excitation–contraction coupling and enhanced contraction (positive inotropic effect)

Time course of membrane conductances g_{Na} and g_K during an action potential
(after Hodgkin and Huxley, 1952)

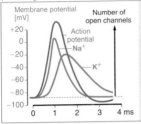

When the threshold is reached, g_{Na} increases due to the opening of many Na channels (see above). As a result of the subsequent inactivation, g_{Na} reaches its maximum before the peak of the action potential and returns to the resting level within 1 ms.

In contrast, g_K increases after a delay and slowly, does not reach its maximum until the middle of the depolarization, and then falls again (because the depolarization is reduced).

Electrotonus and stimulus

Passive, local changes in the membrane potential due to current flow are called electrotonic potentials or electrotonus (ET). Depending on the direction of current flow, the membrane is de- or hyperpolarized (made more positive or negative, respectively; see the figure on p. 10). A depolarization of the resting potential to (or beyond) the threshold constitutes a (suprathreshold) stimulus (see p. 13). In a tissue current originates from activated synapses or from sensory receptors; in experiment and diagnosis (electroneurography, see below) stimulus current is applied by way of electrodes. In all three cases the applied current changes the membrane potential according to the physical properties of the membrane (resistance, capacitance; see below).

Facts and definitions with respect to electrotonus (ET)

Term	Description/comments
Capacitive current	One of the 2 components of current that flows through the cell membrane; is large at the beginning of ET and goes to 0 in the plateau
Resistive (ionic) current	The second component; is 0 at the beginning and reaches a constant maximum in the plateau of the ET
Time constant of the membrane (τ_m)	ET rises and falls exponentially. The rate of change can be characterized by the membrane time constant τ_m, i.e. the time required to rise or fall to 37% = 1/e of the plateau value. Typical τ_m: range 10–50 ms. $\tau_m = R_m \cdot C_m$
Length constant of the membrane (λ)	In elongated cells (axons, muscle fibers) the plateau of the ET falls off exponentially with distance from the site of current flow. The membrane length constant λ is the distance at which the amplitude has fallen to 37% = 1/e. Typical λ are in the range 0.1–5 mm
Local response	An incompletely developed local excitatory state with ET in the near-threshold region; caused by increasing activation of Na channels but not sufficiently for complete excitation
Minimal stimulus current and duration	The current amplitude just adequate to trigger excitation is the minimal stimulus current. The shortest time for which such a current must be applied in order to trigger excitation is the minimal stimulus duration. The minimal duration becomes shorter when the current is stronger but cannot be made arbitrarily short (exploited, e.g., in high-frequency diathermy apparatus)

Conduction of the action potential

Action potential conduction in unmyelinated and myelinated nerve fibers by local currents

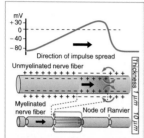

• The all-or-none depolarization during the AP produces a considerable potential difference between the excited and neighboring unexcited cell membrane. This difference causes current flow that depolarizes the adjacent membrane area until the threshold is reached and an AP is generated there, and so on. The AP propagates like a spark along a fuse.

• In unmyelinated nerve fibers the effective current is in the immediate vicinity, producing slow conduction of excitation (as in skeletal, cardiac and smooth muscle fibers).

• In myelinated nerve fibers much of the current flows between nodes of Ranvier, producing rapid saltatory conduction.

• In both types of fibers, the thicker the fiber, the more rapidly the excitation is propagated (see following Tables), because a larger axonal cross-section is associated with lower internal resistance (longitudinal resistance of the axon), so that the electrotonic current flows more rapidly from the excited to the unexcited fiber area.

Supplementary facts and definitions with respect to excitation and conduction

Term	Definition/description/comments
Orthodromic	Designates the normal (physiological) direction of conduction, e.g. from a peripheral sensor (sensory receptor) to the spinal cord, from motoneuron to end-plate, etc. Because every AP pulls a 'train of refractoriness' after it, the AP normally propagates in only one direction
Antidromic	Designates conduction in a direction opposite to the physiological direction, e.g. due to electrical stimulation, pathological excitation. When an axon is stimulated, conduction is both orthodromic and antidromic. If an antidromic AP meets an orthodromic AP, the two extinguish one another
Influence of AP amplitude	The greater the amplitude of the Na^+ influx, the more current is available to charge the neighboring membrane and the more rapidly the threshold is reached there; hence a large AP amplitude causes high conduction velocity. Normally nerve fibers operate in the optimal range
Rhythmic impulse generation	Under prolonged depolarization (as in tonic sensory stimulation) a new AP is generated as soon as the refractory phase has passed, so that trains of impulses are produced with a frequency that depends on the degree of depolarization; in some types of cell groups of impulses ('bursts') may be generated. The various forms of repetitive discharge in the presence of tonic stimulation are produced by K^+ channels with different time and voltage dependence

Classification of nerve fibers (after Erlanger and Gasser compiled by Dudel, 1995)

Fiber	Function, e.g.	Mean diameter	Mean conduction velocity
Aα	Primary muscle spindle afferent, motor axon	15 μm	100 m/s (70–120 m/s)
Aβ	Skin afferent for touch and pressure	8 μm	50 m/s (30–70 m/s)
Aγ	Motor, to muscle spindle	5 μm	20 m/s (15–30 m/s)
Aδ	Skin afferent for temperature and nociception	<3 μm	15 m/s (12–30 m/s)
B	Sympathetic preganglionic	3 μm	7 m/s (3–15 m/s)
C	Skin afferent for nociception, sympathetic postganglionic unmyelinated	1 μm	1 m/s (0.5–2 m/s)

Alternative classification of nerve fibers (after Lloyd and Hunt compiled by Dudel, 1995)

Group	Function, e.g.	Mean diameter	Mean conduction velocity
I	as Aα above	13 μm	75 m/s (70–120 m/s)
II	as Aβ above	9 μm	55 m/s (25–70 m/s)
III	as Aδ above	3 μm	11 m/s (10–25 m/s)
IV	as C above (unmyelinated)	1 μm	1 m/s

Electroneurography (ENG): elicitation and recording of compound action potentials of peripheral nerves to test conduction (modified after Schmidt, 1983)

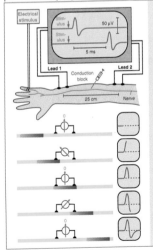

• Extracellular compound action potentials can be recorded from nerves with metal electrodes on the skin surface or needle electrodes stuck into the nerve; they are used in clinical diagnosis to test excitability and conduction velocity of the nerve fibers. The nerves are usually stimulated electrically.

• In the extracellular field only a small fraction (0.1–1%) of the actual AP amplitude is recorded (see calibration in the figure).

• The extracellularly recorded APs also have a different, distinctly biphasic shape. They are produced as shown in the figure: there is always a voltage difference between the recording electrodes (upward or downward deflection of curve) when the wave of excitation, migrating from left to right, is detected by only one electrode.

Definition of synapses, nature of transmission, functions

The sites at which the endings of the axon of a neuron are functionally connected to nerve, muscle or gland cells are called synapses. At synapses, action potentials (and hence the information they contain) may be transmitted to the next cell. Transmission is occasionally direct (electrical synapse), but usually the release of chemical substances called transmitters (see p. 5) is involved (chemical synapse). Activation of a synapse causes either excitation or inhibition of the next (postsynaptic) cell. That is, there are excitatory and inhibitory synapses. Synapses function as rectifiers, are capable of learning (plasticity) and can be modified by drugs.

Neuromuscular junction: prototype of chemical synapses

The 3 main components of the neuromuscular junction

Name	Description
1. Presynaptic ending	The axonal element of the neuromuscular synapse is called the end-plate because it is thickened and broadened; it includes vesicles (diameter ca. 50 nm) that contain the transmitter acetylcholine, ACh (ca. 10 000 ACh molecules per vesicle)
2. Synaptic cleft	Narrow (ca. 50 nm) gap separating the presynaptic from the postsynaptic (subsynaptic) membrane; the zone through which transmitters, enzymes and drugs diffuse
3. Subsynaptic membrane	Membrane of skeletal muscle fiber covered by the end-plate; it is folded to enlarge its surface area and contains ACh receptors that open ion channels (see below) when activated by ACh

Structure and operation of chemical synapses (example: neuromuscular junction)

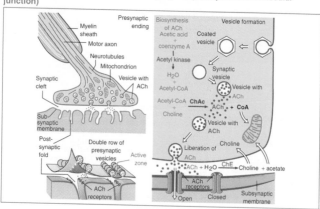

Sequence of 7 main events in the generation of an end-plate potential (EPP), i.e. in neuromuscular transmission

The 7 events described below occur in comparable form at all chemical synapses and are therefore described extensively here and only here. Nerve cells, however, usually have many synapses (often thousands), each of which cannot by itself excite the postsynaptic cell; only when many synapses are activated simultaneously or in rapid succession does the cell become excited (spatial or temporal facilitation, see pp. 24, 25)

Event	Description/comments
1. Presynaptic action potential (AP)	Comes from the motoneuron, propagates along the motor axon into the presynaptic ending and there triggers events 2 and 3 within ca. 1 ms
2. Ca^{2+} influx into presynaptic ending	The depolarization of the presynaptic ending by the incoming AP causes Ca channels to open so that there is an influx of Ca^{2+} ions owing to the potential and concentration gradients (see p. 11). The amount of Ca^{2+} flowing in depends on the degree of depolarization (i.e. the amplitude of the AP)
3. Presynaptic ACh release in 'quanta'	Having entered the presynaptic ending, the Ca^{2+} causes the synchronous exocytotic release of ACh from many synaptic vesicles into the synaptic cleft (1 vesicle releases 1 'quantum' of ACh, ca. 10 000 ACh molecules). At rest single quanta are occasionally released, causing miniature end-plate potentials
4. ACh diffuses to and reacts with the subsynaptic receptors	In fractions of a millisecond the ACh molecules diffuse to the ACh receptors of the postsynaptic membrane and bind to (i.e. activate) them. To each receptor 2 ACh molecules are bound ('cooperative' form of binding). Activation causes the opening of membrane channels permeable to Na^+, K^+ and Ca^{2+} ions
5. Opening of the ion channels allows end-plate current to flow	The end-plate current is produced by the flow of Na^+ and Ca^{2+} ions into and K^+ out of the skeletal muscle fiber (passively, down the diffusion gradients). The current flows for only a short time (1–2 ms). Each individual channel opens and closes abruptly (patch clamp measurements). On average, the result is a rapidly rising and falling current pulse (voltage clamp measurements). Its equilibrium potential is ca. $-10\,mV$
6. End-plate current shifts membrane potential in depolarizing direction: end-plate potential (EPP)	The EPP rises to its maximum in a few ms and then returns to the resting potential with a time constant of ca. 5 ms. The EPP is not actively conducted but rather propagates passively (electrotonically) along the muscle fiber. The normal EPP is always suprathreshold (ca. 30 mV), i.e. it elicits a conducted AP, so that the muscle fiber is excited and twitches
7. Transmitter action is terminated	The enzyme cholinesterase is bound to the basilar lamina near the ACh receptors. It breaks ACh down into choline and acetic acid. After the choline and acetic acid have diffused back into the presynaptic ending, they are used to resynthesize ACh.

Three possible forms of neuromuscular block (neuromuscular relaxation)

The South American arrow poison curare kills by blocking neuromuscular transmission (see below), because the paralysis of the respiratory muscles causes suffocation. Today neuromuscular relaxation accompanied by artificial respiration is an indispensable aid in surgery, to eliminate reflex muscle contraction and relax the musculature despite shallow anesthesia

Event	Remarks
	Block by reduction of presynaptic release (no clinical–therapeutic application)
Withdrawal of Ca^{2+} ions	Reduction of the Ca^{2+} concentration in the bathing (blood substitute) solution reduces the influx of Ca^{2+} into the presynaptic ending on arrival of an AP; not applicable *in vivo* but only *in vitro*
Addition of Mg^{2+} ions	Mg^{2+} ions competitively displace Ca^{2+} ions from the presynaptic Ca channels. Hence the action is like that of Ca^{2+} withdrawal, see above; not applicable *in vivo* but only *in vitro*
Botulinus toxin	Poison produced by botulinus bacteria (spoiled raw meat, ham, preserves); blocks presynaptic transmitter release. Early symptom: drooping eyelids. Heat-sensitive – boil suspect food!
	Antagonistic blockade of ACh receptors (non-depolarizing muscle relaxation)
Curare (*d*-tubocurarine, many synthetic substances, e.g. pancuronium)	Binds to postsynaptic ACh receptors, like ACh, but does not trigger opening of the receptor channels; 'sticks' to the receptors better than ACh and hence displaces it 'competitively', thus acting as antagonist. Curare action is abolished by cholinesterase inhibitors, which allow the concentration of ACh in the synaptic cleft to rise sufficiently to displace the curare
α-Bungarotoxin	Snake toxin that binds irreversibly to ACh receptors. Radioactively labeled, it can be used in experiments to determine the number of ACh receptors
	Block by prolonged activation of ACh receptors (depolarizing muscle relaxation)
Succinylcholine (and others)	Binds to the ACh receptors about as well as curare but opens the receptor channels, and keeps them open longer than ACh because it is only slowly broken down by cholinesterase. The result is maintained end-plate depolarization, inactivation of the neighboring Na channels and hence muscle paralysis; duration of action shorter than that of curare: use in brief anesthesia
Eserine, neostigmine, edrophonium (and others)	Substances that inhibit cholinesterase and thus delay the breakdown of ACh, causing maintained depolarization (consequences as above). Used as antidote to curare overdosage and in myasthenia to improve neuromuscular transmission
Organophosphate insecticides	Many insecticides combine irreversibly with acetylcholinesterase and thereby make it ineffective (consequences as above). Other cholinergic synapses (in autonomic ganglia) are also affected. Some poison gases act by the same principle

Central excitatory chemical synapses

Generalization of the origin of the EPP to other transmitter-induced excitatory postsynaptic potentials (EPSPs), with motoneuron as example

- Every motoneuron has about 6000 somatic and dendritic synapses, some inhibitory and some excitatory. The presynaptic elements in these synapses are called terminal boutons.

- Excitatory postsynaptic potentials, EPSPs, at central neurons such as motoneurons correspond in their origin (events 1–6, p. 19) and time course (see figure below, rise time ca. 2 ms, fall time ca. 10–15 ms) to the end-plate potential. The individual EPSPs have a smaller amplitude, however, so that for suprathreshold excitation many synapses must be activated simultaneously or in rapid succession (spatial and temporal summation, respectively).

- The excitatory transmitter is very probably glutamate (see p. 5). Its action is terminated (event 7) not by enzymatic splitting but by reuptake of the glutamate by active transport from the synaptic cleft into the presynaptic ending (most important step in termination of transmitter action for all classical transmitters except acetylcholine) or into glial cells.

- Site of action potential generation is the axon hillock at the emergence of the axon from the soma, because the threshold of the motoneuron is lowest here.

- The excitability of a motoneuron is greater, the smaller the neuron is. It follows that small motoneurons are more often active than larger ones, and they are easier to recruit during reflexes. This is also true of other neurons.

Monosynaptic EPSP of a motoneuron produced by stimulation of homonymous Ia afferents at increasing intensity (modified after Eccles, 1969)

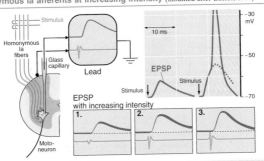

Some of the excitatory synapses on motoneurons are made by the primary afferent nerve fibers of muscle spindles (the Ia fibers, see p. 89) of the neuron's own (homonymous) muscle. The structures involved form the simplest reflex arc in humans, the monosynaptic stretch reflex (myotatic reflex, e.g. patellar tendon reflex, see also p. 92).

Central inhibitory chemical synapses

Origin of inhibitory postsynaptic potentials (IPSPs) with motoneuron as example; similarities to and differences from EPSPs

Inhibitory synapses are analogous to excitatory synapses, except that their activation (events 1–6, p. 19) causes a decreased excitability (inhibition). Postsynaptic inhibition occurs at synapses on the membrane of the inhibited neuron: axosomatic and axodendritic synapses. Presynaptic inhibition occurs at synapses on excitatory presynaptic terminal buttons (axoaxonal synapses) and suppresses the release of transmitter. Inhibitory synapses are just as important as excitatory ones; a blockage of inhibition, as by strychnine or bicuculline (see below), causes convulsions.

Term	Comment
IPSP	Hyperpolarizing potential change, the mirror image of the EPSP with a similar time course (rise ca. 2 ms, fall 10–12 ms). At other inhibitory synapses longer time courses have been observed, and the IPSP can also depolarize by a few mV in some synapses
Ionic mechanism	The reaction between inhibitory transmitter and postsynaptic receptors causes a brief (1–2 ms) conductance increase for K^+ or Cl^- ions, both of which have an equilibrium potential close to the resting potential (see p. 12)
Transmitter	The amino acid glycine is an important inhibitory transmitter (see p. 5); its competitive antagonist is strychnine. Another important inhibitory transmitter, involved in presynaptic inhibition (see below), as well as in IPSPs, is GABA (see p. 5); its competitive antagonist is bicuculline.

IPSP of a motoneuron produced by progressively stronger stimulation of antagonistic Ia afferents (after Eccles, 1964)

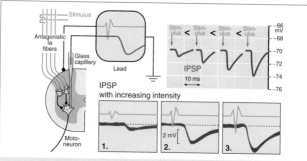

Ia afferents excite inhibitory interneurons, which in turn make inhibitory synapses on the antagonistic motoneurons. This is the shortest inhibitory reflex path (bisynaptic or reciprocal inhibition) in the spinal cord (see p. 93).

The inhibitory action of the IPSP is based on 2 mechanisms

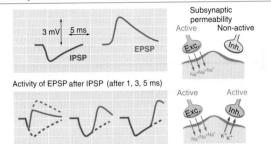

The inhibitory action of the IPSP is based on (1) on hyperpolarization of the membrane potential throughout the IPSP and (2) on an increased membrane conductance while subsynaptic ion currents flow – i.e. a decrease in membrane resistance R so that, according to Ohm's law ($E = R \cdot I$), a given excitatory subsynaptic current I elicits a smaller depolarization E. By means of (1) an EPSP is shifted in the hyperpolarizing direction, 'away from the threshold', and by (2) its amplitude is reduced. Mechanism 2 is the more important for most inhibitory synapses in the CNS.

Operation of presynaptic inhibition

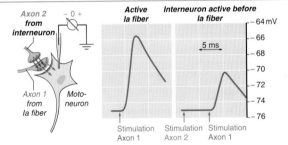

This form of central nervous inhibition operates by the activation of an axoaxonal synapse. The transmitter released from the presynaptic terminal bouton is GABA. It induces an increase in Cl⁻ conductance in the presynaptic terminal bouton (which is postsynaptic to the axoaxonal synapse). The resulting depolarizing synaptic potential causes a reduced transmitter release from the inhibited terminal bouton. The time course is about 100–150 ms, distinctly longer than that of the EPSP. Particular individual synaptic inputs to a neuron can be blocked by presynaptic inhibition without affecting the overall excitability of the cell.

Mechanisms of synaptic interaction and plasticity

Definition of summation, facilitation and occlusion at synapses

Term	Definition/comments
Summation	Addition of excitatory subthreshold synaptic potentials; the interaction of an EPSP and IPSP, as shown in the figure on p. 23 above can also be regarded as summation of potentials with opposite signs
Facilitation	If the combined effect of repetitive stimulation is greater than that of summation alone, facilitation has occurred. Facilitation is thought to result from accumulation of Ca^{2+} in the presynaptic terminals
Occlusion	When each of several different inputs by itself elicits suprathreshold EPSPs in a neuron, simultaneous activation of these inputs may produce no additional postsynaptic excitation (because the system is already saturated). This phenomenon is called occlusion

Summary generalization: When the effect produced by several stimuli presented simultaneously or in rapid succession is greater than the sum of the individual effects, this is called facilitation; if it is smaller, this is called occlusion

Synaptic interactions (with no change in amplitude of individual EPSP and IPSP, i.e. without synaptic plasticity [see next page]) include:

1. Synaptic interactions between EPSPs and IPSPs: see figure on p. 23 and associated text.

2. Spatial summation or facilitation: see figure below and pp. 21, 22

3. Temporal summation or facilitation: EPSPs elicited in rapid succession can summate because of their long time course. As soon as they become suprathreshold, temporal facilitation occurs; i.e. more conducted impulses are generated than would be produced by each single EPSP (see figure below).

Spatial and temporal summation in the nervous system (after Birbaumer and Schmidt, 1996)

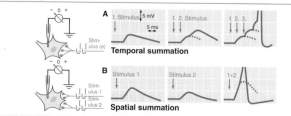

A Temporal summation: Single stimulus (1 arrow) and double stimulus (2 arrows, interval ca. 4 ms) each initiate a subthreshold EPSP; the third stimulus (3 arrows) elicits an action potential. **B** Spatial summation: stimulus 1 and stimulus 2 each initiate a subthreshold EPSP; simultaneous stimulation of both axons (1 + 2) elicits an action potential (truncated at the top in **A** and **B**).

Synaptic plasticity

Synaptic plasticity, a change in synaptic efficiency due to preceding activation, is reflected in the following three processes:
- Synaptic facilitation (tetanic and post-tetanic)
- Synaptic depression (tetanic and post-tetanic)
- Heterosynaptic facilitation

Synaptic facilitation 1: tetanic potentiation (from Dudel, 1990)

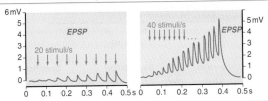

When several presynaptic action potentials arrive at a synapse in rapid succession, the amount of transmitter released is often progressively greater for each successive action potential: tetanic potentiation (in the simplest case, two action potentials, double-pulse facilitation). This form of synaptic plasticity is brought about by 'residual calcium', the elevated Ca^{2+} concentration still remaining in the presynaptic ending after the preceding impulse.

Synaptic facilitation 2: post-tetanic potentiation (PTP) (from Birbaumer and Schmidt, 1996)

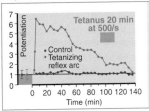

Increase in amplitude of individual EPSP that outlasts or appears after repetitive excitation. From a functional viewpoint this and the above form of synaptic facilitation are central nervous processes made easier by practice and hence reflect a kind of learning. In the hippocampus PTP lasting many hours has been observed. The figure shows PTP of the monosynaptic stretch reflex in the cat.

Synaptic depression

A reduction of the EPSP amplitude in the course of repetitive activation is called tetanic depression. It is the neuronal correlate of fatigue and habituation. A reduction of EPSP amplitude below the control level after the end of repetitive activation is called post-tetanic depression. It has the same functional significance as tetanic depression. In invertebrates it can be demonstrated that the habituation of simple behavioral responses is directly ascribable to post-tetanic depression, and hence represents an elementary learning process just as PTP does.

Heterosynaptic facilitation can lead to long-term potentiation (LTP) (Fig. based on Nicoll *et al.*, 1988)

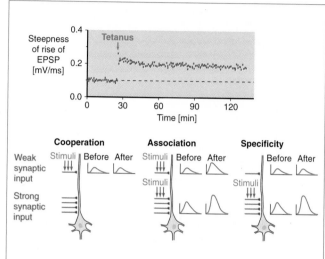

The term heterosynaptic facilitation is applied to facilitation resulting from co-activation of two synaptic inputs, in which activation of one input improves the synaptic efficiency of the second input for a long time (hours or days). This phenomenon is called long-term potentiation (LTP). It is most commonly studied in the hippocampus. LTP is an important participating mechanism in intermediate-term learning processes.

Characteristics of LTP: (1) It appears only when a minimum number of synapses are active co-operatively (the weak synaptic input, above left, does not activate this minimum number, but the strong input, below right, does). (2) It is induced when 2 different inputs are simultaneously active, i.e. it is associative (in contrast to PTP, which always appears in the excited synapses). (3) It is restricted to the excited synapses, i.e. it is specific.

Mechanism of LTP: The excitatory transmitter is glutamate. During normal activation of the synapses it opens only non-NMDA channels (see glutamate receptors, p. 7). During repetitive activation, which causes marked postsynaptic depolarization, NMDA channels are also opened, because Mg^{2+} ions that would otherwise block the NMDA channels are removed when the intracellular milieu becomes more positive. Hence more Ca^{2+} flows into the cells and initiates the processes responsible for LTP (long-term activation of protein kinases as second messengers). As another consequence, a factor released postsynaptically (unknown retrograde messenger) appears to increase the presynaptic release of glutamate.

Electrical synapses

Excitatory electrical synapses

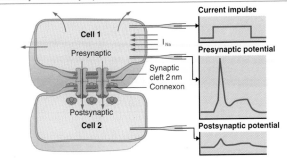

Nexuses (gap junctions) are regions of extremely close contact between membranes (synaptic cleft 2 nm instead of the normal 20 nm) in which half-channels, the connexons, lie exactly opposite one another and together form complete channels from one cell to the other (hence 'junctions across the gap'). Synaptic and action potentials can be conducted, although with poor efficiency, from one cell to an adjacent cell through double connexons. Not only ions but also larger molecules (up to small peptides) pass through the connexons.

The connexons are not constantly open but rather spontaneously adopt open and closed states. They are permanently closed when the pH of the cell falls or the intracellular Ca^{2+} concentration increases greatly (as happens in damaged cells; here closure serves to isolate them from healthy tissue).

Functional syncytia

In various tissues, such as smooth musculature and myocardium, all the cells are linked by nexuses to form functional syncytia. In the myocardium these are the intercalated disks, at which the connections are so close that they can hardly (if at all) be distinguished electrically from the rest of the cytoplasm. Action potentials are conducted in both directions across these cell boundaries. These connections are not synapses in the strict sense; see definition of synapses, p. 18.

Ephaptic transmission

The extracellular currents of an excited cell may, although to a very slight extent, influence the membrane potential of neighboring cells. This form of intercellular communication is called ephaptic interaction. It is normally negligible, but when peripheral nerves and central pathways are injured or diseased, in exceptional cases a suprathreshold ephaptic transmission appears to occur. Such a pathological contact site is called an ephapse. It can be produced by degeneration of the myelin sheath combined with a simultaneous overexcitability of the nerve fibers.

Basic concepts of sensory physiology

Methodological approaches in sensory physiology

• Objective sensory physiology: The investigation of sensory systems with physicochemical methods.

• Subjective sensory physiology (perception psychology): The study of sensations and perceptions with psychological methods.

• Psychophysics: Quantitative measurements relating (objective) stimulus magnitude to (subjective) sensation magnitude.

• Psychophysiology: Simultaneous measurement and linking of objective events to sense organs, or of central nervous phenomena (measured non-invasively, e.g. by EEG) to subjective perceptions and to behavior.

Our sense organs inform us of only a tiny fraction of all that happens in our surroundings and in ourselves (modified after Dudel, 1985)

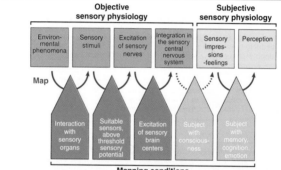

Because we lack suitable sensors, most of the physical and chemical events in the environment and in our own bodies do not act as sensory stimuli; the only events that do so are those for which we have a suitable sense organ (e.g. the ear for sound waves, the eye for light).

The figure shows the processes set in motion by excitation of a sense organ: in the boxes are basic phenomena of sensory physiology, and the arrows between them signify 'leads to by way of' (correspondence or mapping, not direct causality); dashed arrow marks the transition from physiological to psychological processes.

Sensory impressions are the simplest elements of the sensory experience; when several coincide the result is a sensation. When the latter is interpreted by means of our memory it becomes a perception.

Every sensory modality (synonym: sense, modality, i.e. all sensory impressions conveyed by a sense organ, e.g. the eye) has 4 basic dimensions

Dimension	Definition, comments
Quality	Groupings of related/similar sensory impressions, e.g. in vision gray shade (achromatic value) and color
Quantity	Intensity of a sensation, e.g. strength of brightness sensation, saturation of colors
Spatiality	Position of the sensation in the spatial structure of the surroundings and of our body; degree varies in different senses
Temporality	Position of the sensation in the temporal structure of the surroundings; temporal resolution varies greatly in different senses (e.g. very precise in vision, inaccurate in the sense of temperature)

Modality, quantity, quality and their organic substrates, exemplified by the visual system (modified after Dudel, 1985)

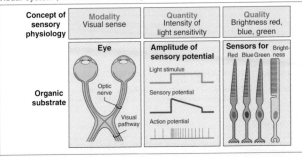

Classification of sensations based on the sensory receptors they employ

Sensors	Modality they mediate
Exteroceptors	Receptors that detect stimuli from the surroundings, e.g. cones and rods in the eye. The five 'classical' senses of sight, hearing, smell, taste and touch employ exteroceptors
Proprioceptors	Receptors that record the position and movement of one's own body, e.g. muscle spindles and tendon organs (see motor system); also the receptors of the vestibular system; together they mediate deep sensibility (proprioception)
Enteroceptors	Receptors that record mechanical and chemical events in the viscera, e.g. in the carotid sinus baroreceptors measure blood pressure and chemoreceptors measure carbonic acid and oxygen tension; enteroceptors mediate visceral sensibility or visceroception. Excitation of enteroceptors produces general sensations (e.g. hunger, thirst), sometimes signaling emergencies (e.g. shortness of breath), but often no consciously perceived sensation

General objective sensory physiology

The specificity of the sense organs is shown by their adequate stimuli and reflected in the 'law of specific sensory energies'

The sensory receptors respond optimally to only one physicochemical form of stimulus (usually the stimulus that excites with minimal energy). This is called the adequate stimulus (e.g. in eye: electromagnetic waves with lengths of 400–800 nm [blue or red]; ear: sound waves at 20–16 000 Hz). But other stimuli, such as electrical, can also excite receptors; they are called non-adequate stimuli. The subjective sensations elicited by both forms of stimulus are always those specific to the sense organ (e.g. for the eye, sensation of light). It is this fact that the 150-year-old 'law of specific sensory energies' basically expresses.

Classification of the sensory receptors according to their adequate stimuli

Type of sensor	Description/comments
Mechanoreceptor	Records mechanical deformation, e.g. in skin, muscle, ear and equilibrium organ
Thermoreceptor	Records cooling or warming, mainly in the skin but also in hypothalamus and other central nervous structures
Chemoreceptor	Responds to chemical stimuli, e.g. the smell and taste receptors but also many enteroreceptors
Photoreceptor	Responds to photons, i.e. to visible light; the rods and cones of the retina
Nociceptor	Is specialized to detect (potentially) tissue-damaging physical or chemical stimuli; nociceptors are present in practically all tissues

Arrival of a stimulus at a sensory receptor initiates the processes of transduction, transformation and conduction

Process	Description/comments
Transduction	Receptors have a normal resting potential. Arrival of stimulus leads to a change in permeability of the receptor membrane (mainly opening of Na channels) and hence to depolarizing ion currents. The resulting depolarization is called a receptor potential. It lasts as long as the stimulus, and its amplitude increases with stimulus intensity; it is an image of the stimulus. The entire process of converting the stimulus into a receptor potential is called transduction
Transformation	Process in which the receptor potential triggers action potentials in adjacent parts of the axon of the afferent nerve fiber. The receptor potential spreads electrotonically into the adjacent parts of the axon, where it depolarizes the resting potential to the threshold for action potential generation. The result is volleys of action potentials, the frequency of which depend on the amplitude of the receptor potential
Conduction	Propagation of the afferent volleys, at a velocity specific to the nerve fiber (see p. 17), to the first synapse in the CNS (spinal cord or brainstem, depending on sensory modality): there the stimulus is recoded into synaptic potentials

Structural, temporal and spatial aspects of stimulus encoding and the coding of stimulus intensity in sensory receptors

Concept	Description/comments
Primary and secondary sensory receptors	In primary sensory cells (the majority) the transformation occurs in the beginning of the axon of the sensory cell. In secondary sensory receptors there is a synapse between an accessory cell and the axon (e.g. hair cells of the inner ear). The receptor potential in a hair cell results in a synaptically mediated generator potential in the axon
Tonic receptor	Represents the amplitude of the stimulus in particular, hence the terms static or proportional sensor; simultaneously measures accurately the duration of the stimulus (good example: Golgi tendon organ: very slight phasic component)
Phasic receptor	Disproportionately large response to changes in stimulus intensity, hence also called dynamic or differential sensor; signals velocity of stimulus change, hence also fairly accurately the stimulus duration. Extreme example: Pacinian corpuscles (vibration receptor, no tonic sensitivity at all)
PD receptor	Receptor with proportional and differential sensitivity; most common receptor type, allowing many gradations and combinations of the two properties. Example of a balanced mixture: primary muscle spindle afferent
Coding of stimulus amplitude	The characteristic function expressing the quantitative relation between stimulus intensity S and tonic discharge frequency F of the afferent burst is usually a power function: $$F = k \cdot (S - S_0)^n$$ where k is a constant and S_0 stands for the threshold stimulus intensity. The exponent n has a characteristic positive value for each receptor type: examples: $n = 1$ for stretch receptors, $n < 1$ for photoreceptors, $n > 1$ for nociceptors
Adaptation	Gradual decrease in afferent volley as stimulus continues unchanged. Phasic receptors are by definition rapidly adapting, but tonic receptors also adapt, though slowly. Exceptions, e.g. nociceptors (prolonged toothache) and cold receptors (feet stay cold for hours). PD receptors are also mixed with respect to adaptation. Threshold (see above) and adaptation protect us from being inundated with trivial stimuli; submechanism of habituation
Primary receptive field	Area, e.g. on the skin, from which a receptor, e.g. a pressure receptor in the skin, can be excited. Many afferent axons branch to supply several receptors, so the primary receptive field can consist of one relatively large or several small areas. In many receptors the primary receptive field cannot be definitely determined, e.g. because higher-intensity stimuli can elicit excitation 'at a distance'. The (secondary) receptive field of central sensory neurons (see below), in contrast, can be unambiguously defined, because it is delimited by central excitatory and inhibitory processes

Mapping and processing in the sensory pathways

A stimulus normally excites many sensory receptors simultaneously. The resulting flood of impulses in the afferents contains all the information about spatial extent, intensity and temporal structure of the stimulus. This information must be evaluated by the CNS. In all senses, evaluation occurs at several relay stations between the sensory receptor and cerebral cortex (e.g. dorsal horn, brainstem, thalamus; details given later). Here are some general principles:

Concept	Description/comments
Divergence	Practically all afferents divide into collaterals after entering the CNS and thus send afferent impulses to several (often many) neurons
Convergence	Practically all sensory neurons in the CNS receive inputs from several (often many) sensory receptors. Divergence and convergence provide a large 'safety factor' in sensory systems, so that loss of a few receptors and neurons is irrelevant
Inhibition	Is needed to prevent unrestricted spread of excitation. Lateral inhibition (surround inhibition) by negative feedback leads to contrast enhancement, descending inhibition blanks out undesired information (e.g. in focusing attention)
Receptive field (secondary)	The entire part of the body periphery from which a sensory neuron can be excited *or inhibited* by stimuli. Often the excitatory receptive field is surrounded by an inhibitory field, and vice versa (contrast enhancement mechanism). The size of receptive fields varies widely and can be altered by inhibitory processes; pathophysiological changes (e.g. due to inflammation) are common
Amplitude coding	The characteristic function for the coding of stimulus amplitude is usually a power function, like that for sensory receptors (see above)
Thresholds	Absolute threshold: smallest stimulus intensity that elicits a detectable change in neuronal impulse frequency; difference threshold: smallest change in a stimulus parameter that causes a change in discharge frequency, e.g. intensity-, position-, timing-, pitch- or color-difference thresholds

Schematic representation of a sensory system (modified after Shepherd 1983)

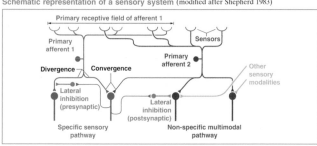

General subjective sensory physiology (perception psychology)

Conscious perception of sensory stimuli usually follows the same rules as information processing in sensory receptors and sensory neurons

In all the senses it is possible to test thresholds, difference thresholds, intensity of sensation, spatial and temporal dimensions as well as the adaptation of subjective sensations and perceptions by using the subject as a 'measuring device'. The subjects indicate their sensations either verbally or non-verbally (e.g. moving a pointer, intermodal intensity comparison). Such measurements can also be made in animals after suitable conditioning, so that the stimulus induces a behavioral change. The following Table summarizes the most important concepts:

Concept	Description/comments
Absolute threshold	Smallest stimulus just capable of producing a sensation; also called stimulus limen, SL. Measured by the method of limits, contrast procedure or signal-detection theory approach
Difference threshold (Weber fraction, jnd)	Synonym: difference limen, DL; just noticeable difference, jnd. The amount by which a stimulus must be made smaller or larger than a comparison stimulus in order to be perceived as just barely weaker or stronger; according to Weber's rule it is usually a constant percentage (Weber fraction) of the comparison stimulus, e.g. 3% (jnd for space and time discussed below)
Weber–Fechner law	Describes sensation intensity E as proportional to the logarithm of stimulus intensity S, i.e. $E \sim \log S$; was long regarded as basic law of psychophysics, but for most senses it applies only to an intermediate intensity range, not to very small or very large stimuli
Stevens' power law	Describes the dependence of sensation intensity E on stimulus intensity S as a power function $E \sim (S - S_0)^n$, just like the coding of stimulus intensity in sensory receptors (see Table above; for most senses applies over a broad intensity range. If the exponent $n < 1$ (normal case), E increases distinctly more slowly than S (just as in logarithmic relation), e.g. when $n = 0.3$, E doubles when S increases tenfold ($10^{0.3} = 1.99$)
Spatial threshold	Smallest distance (jnd, see above) between two stimulus points at which they are perceived as separate, e.g. spatial thresholds for sense of touch, p. 34; acuity for vision, p. 51. Spatial discrimination ability is assisted by contrast enhancement, e.g. in vision the contrast at a boundary between a light and a dark area is perceived as stronger than that corresponding to the physical brightness distribution (simultaneous contrast)
Temporal thresholds, adaptation	Temporal resolution is determined by measuring time difference thresholds (jnd, see above), e.g. by comparing the duration of tones. With periodic stimuli the fusion frequency is measured, e.g. flicker fusion frequency in vision. More prolonged stimuli lead to adaptation, a decrease in sensation intensity (exception: pain)

Components of somatovisceral sensibility, their locations in the body and the basic types of somatovisceral receptors (after Birbaumer and Schmidt, 1996)

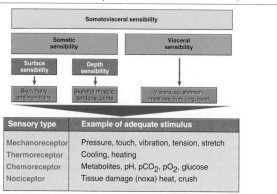

Sensory type	Example of adequate stimulus
Mechanoreceptor	Pressure, touch, vibration, tension, stretch
Thermoreceptor	Cooling, heating
Chemoreceptor	Metabolites, pH, pCO_2, pO_2, glucose
Nociceptor	Tissue damage (noxa) heat, crush

Mechanoreception (sense of touch)

Psychophysically measurable properties of mechanoreception

There are four qualities of cutaneous mechanoreception: the sensations of pressure, touch, vibration and tickle. Each quality is mediated by specific mechanoreceptors (see p. 35), which are distributed over the skin in varying densities. Single mechanoreceptors can be stimulated in isolation with von Frey hairs. The density of such touch points is especially high on the lips and balls of the fingers and especially low on, e.g. the back. Carefully performed measurements of the tactile sense have shown:

Concept	Description/comments
Thresholds of touch	Minimal depth of indentation of the skin for just perceptible touch sensation (tactile sensation threshold) is 0.01 mm; it is lowest on the fingertips
Intensity function	Stevens' power law applies: there are interindividual differences in the exponent n, which is usually around n = 1. Intraindividually n is fairly constant
Spatial thresholds	Simultaneous spatial threshold (2-point threshold) is optimal, ca. 1–3 mm, at the tip of the tongue, lips, fingertips. Successive spatial threshold (tested by placing compass tips one after the other) is distinctly better than simultaneous. Spatial resolution of touch can be improved by practice (the blind)
Thresholds for vibration	Absolute threshold is lowest, 1 µm amplitude, for oscillation frequencies of 150–300 Hz; difference threshold for change in vibration frequency is best < 100 Hz

Structure and position of mechanoreceptors in human skin (after Birbaumer and Schmidt, 1996)

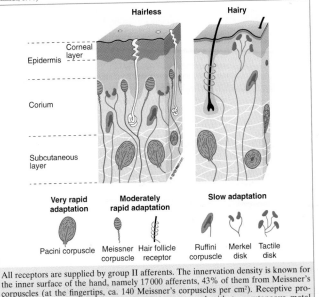

All receptors are supplied by group II afferents. The innervation density is known for the inner surface of the hand, namely 17 000 afferents, 43% of them from Meissner's corpuscles (at the fingertips, ca. 140 Meissner's corpuscles per cm²). Receptive properties of human mechanoreceptors can be measured with transcutaneous metal microelectrodes (transcutaneous microneurography). The associated subjective sensations are recorded at the same time.

Classification of cutaneous mechanoreceptors by their adaptation properties (column headings: SA slowly adapting, RA rapidly adapting) and their adequate stimulus (bottom of columns)

	Adaptation to constant pressure stimulus		
	Slow (SA)	Moderately fast (RA)	Very fast
Hairless skin	Merkel's disks (SA1) Ruffini ending (SA2)	Meissner's corpuscle	Pacinian corpuscle
Hairy skin	Tactile disk Ruffini ending	Hair follicle receptor	Pacinian corpuscle
	Intensity detector	Velocity detector	Acceleration detector
	Classification according to adequate stimulus		

Proprioception (deep sensibility)

Qualities of proprioception

Quality	Description/comments
Sense of position	Informs about the angular positions of the joints and hence about the positions of the limbs with respect to one another and to the head and body (without visual control)
Sense of movement	Informs (also without visual control) about speed and extent of active and passive joint movements. Perception threshold is lower at proximal joints than at distal
Sense of force	Informs about the degree of muscular force necessary to carry out a movement or to maintain a joint position; characterized by great precision and exact reproducibility

Sensory receptors for proprioception (proprioceptors, see also p. 29)

Sensor	Description/comments
Joint receptors	Joint capsules and ligaments contain mechanosensitive sensory corpuscles similar to the Ruffini endings and Pacinian corpuscles in the skin. These signal mainly joint movement
Muscle receptors	Muscle spindles are involved in the senses of position and movement, and also, together with Golgi tendon organs, in the sense of force
Cutaneous receptors	The skin around joints is compressed and stretched during movements. Cutaneous receptors thereby excited could contribute to proprioception, but the contribution is probably small

For proprioception to be perceived, integrative processing of the afferent inputs is necessary (after Birbaumer and Schmidt, 1991)

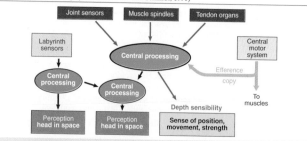

None of the above receptors can mediate deep sensations by itself. The necessary central processing (see Fig.) includes efference copies to remove the inevitable ambiguity of afferent information (e.g. activation of primary muscle spindle afferents either by stretching or by intrafusal contraction). Involvement of the vestibular system is also necessary, to perceive the position of head and body in space.

Thermoreception (sense of temperature)

Temperature sensations of the skin (with the qualities 'sense of warmth' and 'sense of cold')

Static temperature sensations (constant skin temperature)	
Intermediate skin temperature	Warming or cooling in an intermediate temperature range only transiently causes a sensation of warmth or cold. Then complete adaptation occurs. This thermoneutral zone is between 30 and 36°C for small skin areas, and between 33 and 35°C for the whole body unclothed
High skin temperature	Temp. >36°C causes a prolonged warm sensation (more intense, the higher the skin temperature), with a transition to heat pain at 43–45°C; warm sensation is strongest after a temperature step, followed by incomplete adaptation to the long-term sensation
Low skin temperature	Temp. <30°C causes a prolonged cold sensation (more intense, the cooler the skin); onset of cold pain is at 17°C or lower, but the prolonged cold sensation begins to have unpleasant components even at 25°C
Dynamic temperature sensations (during change in skin temperature)	
Initial temperature	Very significant for the resulting sensation. At low skin temperatures the threshold for warm sensation is high and that for cold sensation (or for the sensation 'has become colder') is low; the reverse applies at high skin temperatures
Rate of temp. change	Not important, as long as the change occurs at >0.1°C/s (>6°C/min). With slower temperature changes the warm and cold thresholds increase distinctly and continuously (during very slow cooling one can become 'chilled' without noticing: factor in catching a cold?)
Size of skin area	When small skin areas are stimulated, the thresholds for warm and cold sensation are higher than with large-area heating or cooling (central spatial facilitation of the afferent impulses from the warm or cold receptors)

Warm and cold receptors

Sensor	Description/comments
Warm receptor	Maintained discharge when skin temperature is constant in the range 30–46°C (tonic response,) with maximum at ca. 42°C, plus rise or fall in discharge rate when temperature changes (phasic response, PD receptor); small receptive fields (warm points), group IV afferents (unmyelinated)
Cold receptor	Maintained discharge when skin temperature is constant in the range 20–40°C (tonic response, see Fig.) with maximum at ca. 30°C. Rise or fall in discharge rate when temp. falls or rises, respectively (phasic response, PD receptor): small receptive fields (cold points), group III afferents (thin, myelinated)

The discharges of the thermoreceptors are directly responsible for temperature sensations, though only after considerable central nervous integration (processing) of the afferent inputs (see the complex dependence of warm and cold thresholds on the initial conditions and nature of stimulus).

Dependence of warm and cold thresholds on initial temperature of the human skin (after Kenshalo, 1976)

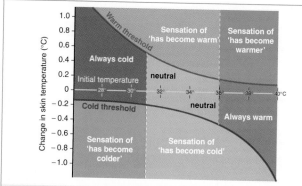

Starting from the temperatures given on the abscissa, to which the skin has been adapted for a long time, the skin temperature must change by the amount indicated (starting from 0) on the ordinate in order to elicit a cold or warm sensation (prerequisite: rate of temperature change >6°C/min).

Behavior of thermoreceptors at constant skin temperature (A) and the response of a cold receptor to short cooling temperature steps (B) ([A] after Kenshalo, 1976, [B] from Darian-Smith *et al.*, 1973)

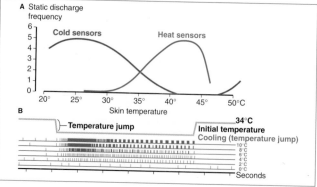

Visceral sensibility

Receptors in the viscera (visceroreceptors) are mainly involved in homeostatic functions (regulatory processes)

Viscera in the thoracic and abdominal cavities contain many receptors, the impulse activity of which is usually not consciously perceived. Instead they send information to the autonomic nervous system, signaling departures of the internal milieu from the desired states (e.g. blood pressure too low or too high, wrong carbonic acid concentration in blood): homeostatic role of visceral receptors. In addition, conscious perceptions may be produced, e.g. pain and other nameable feelings (hunger, thirst, satiety, urge to urinate, shortness of breath, etc.)

Receptors of the viscera

Location	Description/comments
Cardiovascular system	Pressure receptors in aortic arch and carotid sinus measure blood pressure, stretch receptors in the atria of the heart measure filling. However, the heartbeat is perceived by way of mechanoreceptors in somatic structures. Excitation of cardiac nociceptors (due to ischemia) produces heart pain (angina pectoris)
Pulmonary system	Mechanoreceptors in the lung are involved in breathing reflexes, as are chemoreceptors (carotid sinus, brainstem) that measure the carbonic acid and oxygen tension in the blood. Overexcitation of the chemoreceptors produces the feeling of hunger for air and suffocation. Excitation of nociceptors in the airways elicits coughing and sneezing reflexes
Gastrointestinal system	The gastrointestinal tract is part of the body surface. Mechanoreceptors elicit different sensations, depending on their location: satiety when the stomach is stretched, urge to defecate when the rectum is stretched. We are not directly conscious of activity of chemoreceptors in the gut walls (e.g. glucoreceptors, amino acid receptors), though they may contribute to a feeling of satiety. Cold and warm stimuli are perceived only in the esophagus and anal canal. Pain can occur everywhere
Renal system	Mechanoreceptors in the kidneys and ureters do not generate conscious sensations. In the bladder they induce the feeling of need to urinate (depends strongly on attention). Retention of urine in the renal pelvis and ureter (blockage by 'kidney stone') causes severe pain (renal colic). Inflammation of the bladder mucosa produces pain and long-lasting sensation of need to urinate, even when the bladder is empty

Central transfer and processing of somatovisceral information

The thalamus is the only 'entrance gate' to the cerebral cortex. It consists of many nuclei, which can be assigned to 4 classes on the basis of their functions and cortical projection areas

Class	Description/comments
1. Sensory nuclei	Nuclei of the sense organs. Eye: lateral geniculate nucleus; input: optic tract, output: optic radiation to visual cortex. Ear: medial geniculate; input: inferior colliculus, output: to primary auditory cortex. Skin: ventrobasal nucleus, VB; input: lemniscal tract systems (see p. 41), output: to postcentral gyrus (primary somatosensory cortex, SI, and the secondary somatosensory cortex lateral to it, SII)
	The ventrobasal nucleus (VB) has two main divisions: the ventral posterolateral nucleus (VPL, receives projections from the body) and ventral posteromedial nucleus (VPM, receives projections from the face). Inputs are the corresponding lemniscal tract system (q.v.)
2. Non-specific nuclei	Various, mainly medially situated nuclei with no clearly delimitable cortical projections. Inputs: ascending extralemniscal tract systems (see p. 41), outputs: diffuse projections to practically all cortical areas; part of the ARAS (see p. 120)
3. Motor nuclei	Various nuclei with predominantly motor function, e.g. ventral lateral nucleus (VL); inputs: basal ganglia and cerebellum, output: to primary motor cortex and premotor cortex (precentral gyrus)
4. Associative nuclei	Various nuclei with integrative functions, e.g. medial dorsal nucleus (MD) projects to frontal associative cortex

Thalamus of the right half of the brain with its ipsilateral projections to the cerebral cortex (from Zimmermann, 1990)

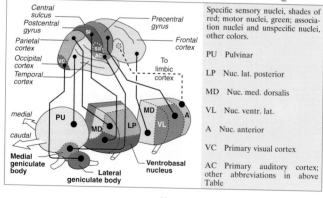

Specific sensory nuclei, shades of red; motor nuclei, green; association nuclei and unspecific nuclei, other colors.

PU   Pulvinar

LP   Nuc. lat. posterior

MD   Nuc. med. dorsalis

VL   Nuc. ventr. lat.

A   Nuc. anterior

VC   Primary visual cortex

AC   Primary auditory cortex; other abbreviations in above Table

Ascending spinal and supraspinal somatovisceral tract systems (Willis and Coggeshall, 1991)

Specific lemniscal tract systems	
Dorsal column path	Group II afferents from mechanoreceptors of the trunk and the limbs. Synapses in dorsal column nuclei (gracilis and cuneate nuclei); from there continues as medial lemniscus after crossing to opposite side, to ventrobasal nucleus of the thalamus and on to the postcentral gyrus
Spinothalamic tract	Part of the anterolateral funiculus with afferents from thermoreceptors and nociceptors. Axons decussate segmentally and in the brainstem join the medial lemniscus
Trigeminal nerve (V)	Afferents from the face and mouth region. Main sensory nucleus is the relay station for mechanoreceptors. Axons decussate and join the medial lemniscus, as do some of the axons from the spinal trigeminal nucleus with information from thermoreceptors and nociceptors
Unspecific extralemniscal tract systems	
Spinoreticular tract	Part of the anterolateral funiculus with afferents from thermoreceptors and nociceptors. Axons decussate segmentally and end in the reticular formation; pathway continues from there to the medial thalamic nuclei
Spinothalamic tract	Part of the anterolateral funiculus with afferents from thermoreceptors and nociceptors. Axons decussate segmentally and run to the medial thalamic nuclei
Trigeminal nerve (V)	Majority of the axons from the spinal trigeminal nucleus, from thermoreceptors and nociceptors in the face and mouth region; end in reticular formation; pathway continues from there to the medial thalamic nuclei

The primary sensory cortex, SI, is organized somatotopically, i.e. there is an orderly spatial mapping of the body surface onto the (contralateral) cortical surface

• Somatotopy has been illustrated in the form of a sensory homunculus. It illustrates the fact that the body areas especially important for tactile sensations (mouth region, fingertips) occupy disproportionately large areas on the SI: a particularly large part of the central apparatus is made available to the dense sensor systems in the periphery, for optimal evaluation of the information they provide (for motor cortex see p. 96).

• The secondary sensory cortex, SII, is also somatotopically organized. Here, however, both halves of the body are mapped onto each hemisphere (bilateral projection); it is probably responsible for bilateral co-ordination of sensory and motor functions (e.g. two-handed activities).

• Conscious somatosensory perceptions ordered in space and time are possible only with an intact cerebral cortex. Furthermore, we must be awake and must direct our attention to the event to be perceived. These processes involve primarily the unspecific sensory system (key words: extralemniscal tracts, reticular formation, ARAS, arousal).

Course of the somatosensory pathways, with the most important relay stations
(after Schmidt, 1983)

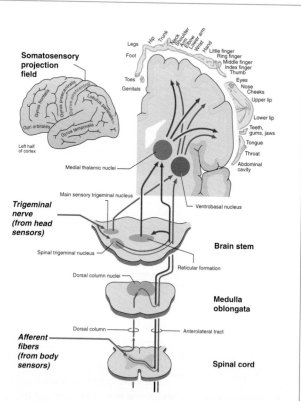

Specific pathways are drawn in red. non-specific in blue. The position of the postcentral gyrus (somatosensory projection field SI) can be seen in the side view of the brain, above left. The schematic drawing of Penfield and Rasamussen, showing the topographically arranged projection of the body periphery onto the postcentral gyrus (sensory homunculus), is included to illustrate the relative sizes of the projections of individual parts of the body onto the cerebral cortex.

Centrifugal control of the afferent influx in the somatosensory system is an example of descending inhibition in all sensory systems; important function: protection against inundation with unimportant information (after Zimmermann, 1990)

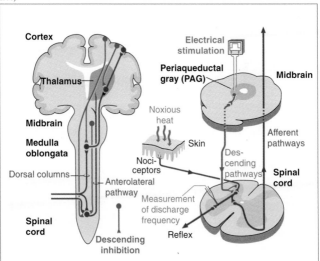

The diagram on the left summarizes the somatosensory descending inhibitory systems (drawn in red). The right diagram shows how afferent information from cutaneous receptors can be inhibited in the spinal cord by activation of the PAG in the midbrain (submechanism of opiate action, because opiates excite PAG neurons).

Pathophysiological disturbances of somatosensory function can manifest as

- Sensory deficit symptoms (e.g. hypesthesia or dysesthesia following damage in the lemniscal system).

- Symptoms of stimulation (e.g. complex paresthesias caused by ectopic impulse generation when ascending pathways have been damaged).

- Disturbances of sensory discrimination functions (e.g. astereognosis, disturbances of body scheme, hemineglect).

There are many definitions of pain: none is universally accepted; a useful one follows

> Pain: An unpleasant sensory and emotional experience associated with actual or potential tissue damage, or described in terms of such damage.

Characterization of pain

Pain has various qualities, which are closely correlated with its site of origin
(from Schmidt, 1990a)

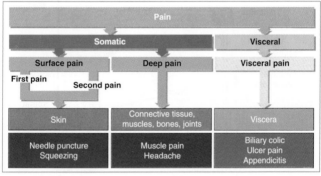

The distinctions between acute and chronic, as well as organic versus psychogenic pain, are as follows

Kind of pain	Definition/comments
Acute	Organically induced pain of brief duration; limited to the site of damage; often definitely localizable; pain intensity proportional to stimulus intensity, pain ends after end of stimulus; clear signal and warning function, activation of autonomic and motor systems
Chronic	Organically induced pain that lasts a long time (>1–3 months) or returns repeatedly (e.g. migraine, trigeminal neuralgia); pain intensity often greater or less than stimulus intensity; autonomic changes, affective elements superimposed. Chronic pain may persist after removal of the organic cause or reappear (with no new organic damage).
Psychogenic	(Synonym: psychologically induced). Pain as direct consequence of social circumstances, emotional events or mental disease (with no organic stage); is experienced like organic pain (see definition of pain, '. . . or described in terms of such damage') Example of psychogenic pain; no physiological but often a social function (e.g. pension claim)

A person's evaluation of experienced pain (cognitive component) and the resulting expressions of pain (psychomotor component) involve sensory, affective, autonomic and motor components (after Schmidt, 1990a)

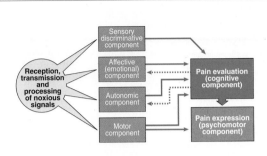

A crucial factor in the evaluation of pain is the comparison of present pain with pain in the past and the consequences at that time. This cognitive evaluation in turn influences the magnitude of the affective and autonomic components (dashed arrows). Other common influences on the evaluation of pain: social situation, family background (upbringing), ethnic origin, circumstances in which pain arises (e.g. accident, war wound, tumor)

Experimental algesimetry: psychophysical measurements (methods, p. 33) of the relationships between noxious stimulus (mechanical, thermal, electrical or chemical 'pain stimulus') and pain sensation

- Subjective algesimetry: measurement of pain threshold, pain intensity (indicated verbally or non-verbally), pain tolerance threshold (stimulus intensity at which the subject asks for the stimulus to be stopped), pain adaptation (usually absent)

- Objective algesimetry: measurement of motor (e.g. flexor reflex, EMG) and autonomic (e.g. pupil diameter) reactions, event-related potentials (ERP), cerebral blood flow (PET, MRI), etc.

- Multidimensional algesimetry: combination of methods of subjective and objective algesimetry, e.g. subjective: pain intensity, objective: pupil diameter, skin resistance (galvanic skin response)

Clinical algesimetry

Application of the methods of experimental algesimetry to the patient; in addition, use of pain questionnaires (e.g. McGill Pain Questionaire) and comparison of the intensity of clinical pain with that of pain imposed experimentally (e.g. tourniquet pain quotient)

Neurophysiology of pain (nociception)

Definitions (see Willis, 1985)

• Nociception: transduction, transformation, conduction and central nervous system processing of noxious (tissue-damaging or potentially tissue-damaging) stimuli.
• Nociceptive system: the peripheral and central neural structures involved in nociception. The subjective sensation of pain is a common, but not obligatory consequence of activity in the nociceptive system.
• Specificity theory of pain: the concept that the nociceptive system represents the function-specific sensory channel for the sense of pain (in analogy to the visual system for seeing, auditory system for hearing, etc.) has strong experimental support. Alternative views with little experimental support: intensity theory (postulates that every somatovisceral receptor can encode noxious stimuli, by a high discharge rate) and pattern theory of pain (postulates the encoding of noxious stimuli by special discharge patterns of somatovisceral receptors).

Elements of the nociceptive system, their properties and functions

Element	Properties, functions, comments
Nociceptor	Sensory receptor with high threshold, usually sensitive to high-intensity mechanical as well as chemical and thermal stimuli (noxae), i.e. polymodal; non-corpuscular ('free') nerve ending, present in practically all body tissues; in normal tissue partially inexcitable ('sleeping'). Function: transduction and transformation (p. 30) of noxious stimuli; sensitivity altered by sensitization (e.g. by inflammation mediators) and desensitization (e.g. by analgesics)
Afferent nerve fiber	Either thin myelinated (group III, synonym: Aδ) or unmyelinated (group IV, synonym: C fiber, p. 17). Function: conduction (p. 30). Distinctly more group IV than group III fibers present. Transmission of initial pain by group III, of delayed pain by group IV fibers. Local anesthetics (mechanism p. 14) block conduction of nociceptive impulses (e.g. in dental medicine) but also other sensory information (hence the region innervated by the blocked nerve is not only analgesic but also anesthetized)
Ascending tract systems	Nociceptive afferents end in the dorsal horn of the spinal cord or the corresponding trigeminal nucleus. Connection there with ascending tract systems that convey the noxious information to the thalamus and cortex (description and illustration of the tracts pp. 41, 42; of the thalamic nuclei p. 40; of the sensory cortex p. 41). Function: central conduction and processing of noxious information. The generation of conscious pain sensations involves the somatosensory cortex and associative cortical areas
Descending tract systems	These have a predominantly inhibitory function in the nociceptive system; hence are described under 'endogenous pain control systems' on p. 48

Pathophysiology of nociception and pain

Definitions of pathophysiological pain sensations

Allodynia	Pain caused by non-noxious stimulation of normal skin. This expression designates conditions in which a normal mechanical or thermal stimulus to the skin elicits pain sensations: the receptors involved may be mechanoreceptors or thermoreceptors; depends on altered central processing
Anesthesia	Abolition of all cutaneous sensory modalities
Analgesia	Absence of pain during noxious stimulation
Dysesthesia	Unpleasant abnormal sensation, either spontaneous or elicited by a stimulus
Hypesthesia	A reduced sensitivity to somatosensory stimuli. The type of hypesthesia should be specified with respect to region and to modality or stimulus form
Hypalgesia	Reduced sensitivity to noxious stimuli. Hypalgesia is usually a component of hypesthesia (see above)
Hyperesthesia	Increased sensitivity to non-noxious stimuli. The type of hyperesthesia should be specified with respect to region and stimulus form
Hyperalgesia	Increased sensitivity to noxious stimuli. It involves a lowering of the threshold to noxious stimuli and/or an intensification of the response to a noxious stimulus
Hyperpathia	Pain syndrome characterized by delayed onset, increased response and an after-response, which outlasts the stimulus. It is most clearly apparent during repetitive stimulation. Hyperpathia can be combined with hypo-, hyper- or dysesthesia
Paresthesia	Abnormal but not unpleasant sensation, either spontaneous or stimulus-induced

Projected (neuralgic) pain is produced by pathophysiological impulse generation in the nociceptive afferent nerve fibers (after Schmidt, 1990a)

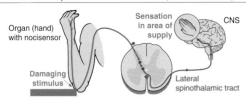

Acute impulse generation in afferents produces a brief dysesthesia (e.g. stimulation of ulnar afferents when elbow is struck); chronic impulse generation in nociceptive afferents leads to neuralgic pain, restricted to the region supplied by the affected nerve or the affected dorsal root (e.g. compression of a spinal nerve in a disk syndrome). Neuralgic pain often recurs in waves or attacks.

Referred pain

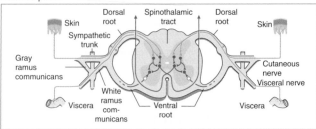

Disorders of internal organs often produce sensations of pain on the body surface: referred pain (each organ is associated with typical areas on the surface, e.g. inner side of left arm for heart). Caused mainly by convergence of visceral and somatic nociceptive afferents onto dorsal horn neurons of the nociceptive system (left): in addition, some nociceptive afferents supply both superficial and deep tissue (right)

Central pain

Pain that results from overexcitability or pathological spontaneous activity in the nociceptive system. Examples: anesthesia dolorosa (after dorsal roots have been torn out), phantom pain (after amputations), thalamic pain (associated with diseases of the sensory ventral nuclei of the thalamus).

Endogenous and exogenous inhibition of pain

Pain control systems within the body (endogenous inhibition)

System	Definition/action/comments
Endogenous opiates	At many places in the nociceptive system there are opiate receptors (four subtypes known at present). Under certain conditions (e.g. extreme stress situations) the body releases endogenous opiates, which bind as ligands to these opiate receptors and induce analgesia. Important examples: polypeptide dynorphin, contains the pentapeptide leu-enkephalin; polypeptide β-endorphin, contains the pentapeptide met-enkephalin
Descending inhibitory systems	From the periaqueductal gray matter (PAG) and the adjacent lateral reticular formation, tracts descend to synapse in the nucleus raphe magnus (NRM) and locus coeruleus. Descending projections terminate in the dorsal horn of the spinal cord and inhibit nociceptive neurons by release of monoamine transmitters (serotonin, norepinephrine); see gate control theory. The PAG contains very many opiate receptors and is an important target of morphine

Gate control theory: Postulates that nociceptive information can be modulated at as low a level as the dorsal horn, by afferent and descending inhibition. Details of the mechanism are unknown. Therapeutic attempts to activate this control mechanism in order to inhibit pain (e.g. by electroacupuncture, stimulating the dorsal column with implanted electrodes, and transcutaneous nerve stimulation, TNS) have so far had only limited success.

Target sites for pain therapy

Procedure	Action/comments
Non-narcotic analgesics	Substances that inhibit pain with no restriction of consciousness (narcosis); many groups of substances and preparations (example: acetylsalicylic acid); predominant target: peripheral nociceptors
Narcotic analgesics	Substances with a strong analgesic action which also limit consciousness, especially in high doses (oldest example: morphine, contained in opium; today many opiates [syn.: opioids] are in use); activate endogenous pain control systems by binding to opiate receptors
Psychotropic drugs	No direct analgesic action, but can have a favorable effect on emotional components of the pain experience. Main groups: tranquilizers and antidepressants
Local anesthetics	Block the generation (surface anesthesia) or conduction of impulses in nociceptive (and other) afferents (nerve block)
Physical measures	Depending on the origin of the pain, a great variety of methods are employed, ranging from application of heat and cold to electrical stimulation and neurosurgery, and also including various forms of therapeutic exercise. In neurosurgery the only procedure of some significance is cordotomy (transection of the anterolateral funiculus of the spinal cord), to interrupt conduction of nociceptive signals from the contralateral half of the body
Psychological methods	Used especially but not exclusively for pain with no clear peripheral cause. Typical methods: biofeedback, operant conditioning, relaxation, meditation, hypnosis

Itch

Is itch a form of pain?

Not enough is known about the pathophysiological sensation itch, which is related to pain. It occurs only in the skin and the adjacent mucous membranes, where it is triggered by the release of histamine. The main open question is whether it is an independent sensation (with its own receptors, etc.) or a special form of the pain sensation.

The eye and its optical system (dioptric apparatus)

Refractive power is measured in diopters (D), dimension m^{-1}

The shorter the focal length f of a convex lens, the larger is its refractive power, RP. Hence the unit used for refractive power RP is the reciprocal of f in meters, the diopter, D. Examples: f = 4 m gives RP = 1/4 = 0.25 D; f = 0.5 m gives RP = 1/0.5 = 2 D; f = 0.25 m gives RP = 4 D. Concave lenses have negative diopters. The eye has various refractive surfaces, the properties of which can be combined in a simple optical system, the reduced eye: focal length in air 17 mm in front of the eye, 23 mm behind it; the overall refractive power of the eye at rest (unaccommodated) is 58.6 D.

The optical system of the eye casts onto the retina an inverted and much reduced image of the surroundings. The amount of incident light is controlled by pupillary responses, the sharpness of the image by accommodation. These involve:

Element	Function(s)/description/comments
Cornea	Anterior, transparent part of the sclera enclosing the eye, ca. 1 mm thick, with no blood vessels; covered by the conjunctiva and continuous with the sclera, which looks white and contains blood vessels; outer surface kept moist and sterile by lacrimal fluid, inner bounded by anterior chamber of the eye. Sensory innervation by first branch of trigeminal nerve (blink reflex when touched). Overall RP 43 D, unalterable. Distance between anterior corneal surface and retina: 24.4 mm
Lens	Biconvex, elastic, transparent connective tissue body surrounded by lens capsule, which is connected to ciliary muscle (see below) and sclera by zonule fibers. Intraocular pressure pulls lens flat. Refractive power at rest 19.1 D (flattened state, unaccommodated). Lens plus cornea, aqueous humor in anterior chamber and vitreous body constitute the compound optical system of the eye, the dioptric apparatus; overall RP 58.6 D, see above
Iris with pupil	Flat disk of tissue in front of the lens with circular central opening, the pupil. Iris contains the smooth dilator (sympathetic innervation) and sphincter (parasympathetic innervation) muscles of pupil, which regulate pupil diameter and hence amount of light entering the eye (extremes: mydriasis and miosis). Light response always simultaneous in both pupils ('consensual'), even with light into only one eye. Pupil also narrows during accommodation: near-vision response. Degree of iris pigmentation determines eye color, from blue to gray to brown
Ciliary muscle	Annular smooth muscle around the lens, parasympathetic innervation (oculomotor nerve). Its contraction relaxes the zonule fibers, so that the lens curvature increases, especially the front surface: accommodation (front and back focal lengths shorten). Accommodation range (max. increase in RP): 14 D at age 10 (near point 7 cm), 2 D at age 50, 0.5 D at age 70. (Presbyopia caused by loss of lens elasticity; near point moves away from eye: loss of RP is compensated by reading glasses)
Retina	Light-sensitive lining of back inner surface of eye; contains the photoreceptive cones (color vision) and rods (twilight vision); also a network of higher-order nerve cells, the final layer of which comprises the ganglion cells. In the visual axis is the fovea centralis (site of most acute vision, cones only)

Subjective and objective methods for eye examination

Method	Definition/description/comments
Acuity measurement	Subjective measurement of visual acuity at the site of greatest acuity (fixation point in optic axis, is projected onto fovea centralis) with test charts (Landolt rings, picture or letter charts). The result is usually expressed as a Snellen fraction (ratio of the distance at which a symbol is discriminated to the distance at which it subtends 1 minute of arc, in practice usually 20 feet; normal: 20/20)
Perimetry	Measurement of visual field of each eye with perimeter apparatus; tested with white and colored light spots (visual field larger for white than for colored). Visual field deficits are called scotomata. A physiological scotoma is the blind spot, where the optic nerve enters the retina; ordinarily not noticed because filled in perceptually. Visual fields of the two eyes overlap only partially, hence binocular field > monocular. Binocular visual field is expanded by eye movements = 'field of gaze')
Ophthalmoscope	Used to study the fundus of the eye by way of the light it reflects. Depending on the method used, the image is upright or inverted (enlarged 16- and 4-fold, respectively). Refractive anomalies found in the eye must be corrected with lenses (objective measurement)
Tonometry	Measurement of intraocular pressure (by applying pressure to the cornea); normally 15–16 (limits 10 and 21) mmHg; produced by ultrafiltration of plasma from capillaries in the ciliary body to form aqueous humor ($2\,mm^3$/min), which flows from posterior to anterior chamber and thence through the canal of Schlemm into the venous system. Blockage of outflow raises the pressure (glaucoma), which damages the retina and may cause blindness
Measurement of dark adaptation	Color vision by day, photopic vision (with cones), gives way to black–white scotopic vision (with rods) as brightness decreases, along with a rapid deterioration of acuity. Because the fovea has only cones, it is impossible to fixate at night (hence scotopic vision has 2 blind spots). Light-adapted eye requires ca. 30 min to adjust to twilight; dark-adapted eye is at first dazzled by sudden brightness but then quickly adjusts. Night blindness, nyctalopia, can be congenital or caused by vitamin A deficiency
ERG	Electroretinogram, ERG, recordable with macroelectrodes from outside of eye. Illuminating or darkening the eye elicits a characteristic sequence of voltage fluctuations (waves a, b, c, d), caused by excitation and synaptic transmission in the retina. Across the resting eye is a steady potential, the cornea being positive with respect to the retina
EOG	Electro-oculogram, EOG, employs the steady corneoretinal potential (forms electric dipole with surrounding electric field) to measure eye movements
VEP	Visual evoked potential, can be recorded over the occipital cortex after light stimuli; complex sequence of waves, strongly dependent on stimulus form (light flash, checkerboard pattern, color, etc.), see p. 114

Structure of the eye: important examination methods

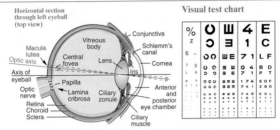

Horizontal section through left eyeball (top view)

Visual test chart

Perimetry

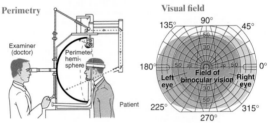

Visual field

Examiner (doctor)

Perimeter hemisphere

Patient

Field of binocular vision — Left eye / Right eye

Eye mirroring in reverse picture and upright picture

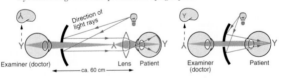

Direction of light rays

Examiner (doctor) — Lens — Patient
ca. 60 cm

Examiner (doctor) — Patient

Dark adaptation

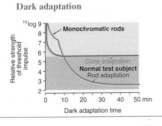

Monochromatic rods
Cone adaptation
Normal test subject
Rod adaptation

Relative strength of threshold impulse — $^{10}\log$
Dark adaptation time

EOG

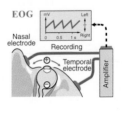

Nasal electrode
Recording
Temporal electrode
Amplifier

Optical defects of the eye, refractive anomalies, spectacles and contact lenses

Term	Description/comments
Spherical aberration	The dioptric apparatus refracts peripheral rays more strongly than those near the optic axis. The resulting blurring is reduced by constricting the pupils (excluding peripheral rays)
Chromatic aberration	Short-wavelength (blue) light is more strongly refracted than long (red) in the dioptric apparatus. Hence greater accommodation is required for red objects at a given distance, i.e. blue objects appear further away (often evident in viewing stained glass church windows, e.g. Chartres)
Astigmatism	The corneal curvature is not ideally spherical but (usually) stronger vertically than horizontally: a dot is imaged as a stripe: up to 0.5 D physiological; when greater is corrected by cylindrical glasses or correspondingly ground contact lenses
Myopia (nearsightedness)	Eye too long relative to refractive power of the dioptric apparatus. Image plane is in front of retina, hence blurred image of distant objects. Far point (definition: acute vision without accommodation) is too close (hence 'near' sighted). Near point (shortest distance for acute vision with maximal accommodation) little changed. Correction: concave lens (negative diopters) or corresponding contact lenses
Hypermetropia (farsightedness)	Eye too short relative to refractive power of the dioptric apparatus. Image plane behind retina, hence acute vision impossible without accommodation (no far point). Because accommodation is needed to see at a distance, near point moves away from eye (hence 'far' sighted). Correction: convex lens (positive diopters) or corresponding contact lenses
Squint (strabismus)	Accommodation is accompanied by convergence of the optic axes, to keep the image of the fixated object on the fovea centralis; in hypermetropia this necessarily causes squint, because the optic axes converge (instead of remaining parallel) even when looking into the distance. Correct glasses (or contact lenses) prevent squint. Squint can also have other causes (e.g. congenital); surgery may help
Presbyopia	Associated with aging. Loss of lens elasticity in older people reduces the range of accommodation (see ciliary muscle in preceding Table). Far point unchanged. Near point moves away from eye. Correction: convex lens for near vision (reading glasses)
Cataract	Clouding of the lens in old age, mainly caused by water incorporation and the development of fissures. Vision restored by surgical removal and glasses with strong convex lens (ca. 13 D for distance vision); it is also possible to insert a plastic lens during the operation

Tears

- Protect the eye from drying out, together with the mucus secreted by goblet cells in the conjunctiva.
- The lacrimal glands produce about 1 ml of tear fluid per day in each eye; tastes salty, is slightly hypertonic.
- Tear secretion can be increased by foreign bodies or by emotions (parasympathetic innervation of pterygopalatine ganglion).

Ray path in emmetropia (normal vision), myopia and hyperopia

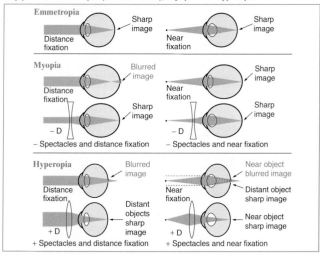

Psychophysiology of vision
Vision provides the basis for perception

	The eye does not send an unambiguous image of the surroundings to the brain. The brain must interpret the impulses arriving from the optic nerves on the basis of its experience, so that we do not see meaningless patterns of visual stimuli but rather perceive objects in the environment. The following aspects are important here:
Size constancy	Image size on the retina is halved for each doubling of the distance of the viewed object. Nevertheless, it is always seen as about the same size
Shape constancy	People or objects known to us are always recognized as the same, regardless of viewing conditions (light, perspective distortion, distance, etc.). The brain employs many mechanisms and kinds of information (e.g. contours, contour intersections or interruptions, horizontal disparity) to arrive at the perception of a closed shape ('Gestalt')
Optical illusions	The data provided by the retina are often ambiguous, as in Necker cubes and other 'impossible' figures. Then the interpretation switches back and forth between alternative solutions. Sometimes misinterpretations can occur, because normally reliable evidence is misleading in a special case. The result is a so-called optical illusion: see Fig. p. 56

Special features of light–dark perception

Concept	Definition/description/comments
Eigengrau	After a long time in darkness a medium gray (Eigengrau = 'intrinsic gray') is perceived, often together with lighter 'nebulae', dots of light, etc., probably due to spontaneous activity in the visual system
Shades of gray	In daylight 30–40 different shades of gray can be discriminated, from deepest black to lightest white. Interpretation (perception) always depends on the surroundings: black checkerboard squares immediately look dark when light goes on in a dark room (even though the retina is receiving more light from the black squares than under Eigengrau conditions)
Simultaneous contrast	The visual system emphasizes (enhances) contours and contrasts. Hence a gray field seems lighter in dark surroundings than in light, etc.: simultaneous contrast; particularly clear along a light–dark boundary: simultaneous border contrast (Mach bands)
Afterimages	Consequence of local adaptation of the retina by preceding illumination. In addition to light–dark afterimages, they are often colored (in the appropriate opponent color red/green, blue/yellow)
Flicker fusion frequency	Light stimuli can be resolved at repetition rates up to 30/s, above which the light appears continuous; the critical flicker frequency is higher at higher light intensity
Phi phenomenon	The term for apparent movement, such as that perceived when lights in a row are consecutively turned on and off

Special features and advantages of binocular vision

Concept	Definition/description comments
Visual field	Definition on p. 51 (perimetry). In binocular vision this, like the field of gaze, is larger than in monocular. There is a considerable region of overlap in the middle, called the binocular field. Lateral regions are seen by only one eye because of the nose
Convergence	Serves to measure distance. The closer a fixated point comes, the more the optic axes converge. The convergence angle can be evaluated by the brain as a measure of the distance of the fixated object (principle of the split-image range finder)
Horizontal disparity	Each eye sees a given object from a different horizontal position: horizontal disparity. As a result, objects are imaged on the retina in such a way that all objects closer than the fixation point would have to appear as crossed double images, and all those further away as uncrossed ones. The brain processes this information to produce binocular fusion, generating an impression of spatial depth; effective only up to a distance of about 6 m. When the complex interplay of convergence and disparity is disturbed (e.g. slight displacement of an eyeball with a finger, eye muscle paralysis), binocular fusion disintegrates and double images appear

Sensory illusions in form perception, ambiguous and 'impossible' figures, example of simultaneous contrast

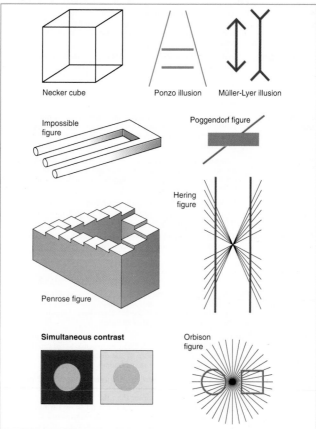

Necker cube

Ponzo illusion

Müller-Lyer illusion

Impossible figure

Poggendorf figure

Hering figure

Penrose figure

Simultaneous contrast

Orbison figure

Color vision

Physical aspects of color vision, measurement of color

Concept	Definition/description/comments
Spectral color	Due to chromatic aberration (see p. 53) a prism separates white sunlight into the monochromatic spectral colors red, orange, yellow, green, blue, indigo and violet, with electromagnetic wavelengths of 700–400 nm
Mixed color	General term for all non-monochromatic spectral colors. Mixing of red and blue gives purple hues that are not present in the spectrum. However, all the spectral colors can also be produced by mixing (see below)
Achromatic valencies	The set of shades of gray from the most radiant white to the deepest black. Saturation of a color is determined by its achromatic component. Furthermore, color valencies are produced that cannot be produced by mixing spectral colors: spectral red plus white gives pink, plus black gives brown
Additive color mixtures	Produced when light of different wavelengths from self-luminous sources falls onto the same place on the retina, e.g. green and red mix in this way to give yellow, see Fig. (Yellow also exists as a spectral color, see above.) For every luminous light source in the color circle there is another that can be additively mixed with it to give white: complementary colors
Subtractive color mixtures	Colors can be derived from white light by using color filters, e.g. a broad-band blue filter also lets green through but no red and a broad-band yellow filter also passes green but no blue: what remains is green

Spectral colors obtained by splitting sunlight with a prism (see above) and diagram of additive and subtractive color mixtures (color mixture illustrations from Grüsser and Grüsser-Cornehls, 1990)

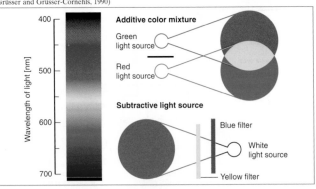

Trichromatic color vision, color theories

Theory	Description/comments
Trichromatic theory	Finding: Three hues suffice to mix all colors. Various combinations of three primary colors can be selected from the spectral colors. By international convention: 700 nm (red), 546 nm (green), 435 nm (blue). Theory: color vision is based on differential activation and subsequent common processing of excitation in these three color systems. Neurophysiology: the retina contains three cone types with different absorption maxima, i.e. different visual pigments
Opponent process theory	Finding: intense red stimulus produces green afterimage (and vice versa); the same holds for blue/yellow and white/black. Theory: color vision is based on four primary colors that act as antagonistic pairs, the opponent colors red/green, blue/yellow. Neurophysiology: in the retinal neurons on which cones synapse, antagonistic excitatory and inhibitory processes are elicited by the opponent colors, and an additional black–white system is present as a brightness system. That is, both theories are 'correct' at different levels of the visual system

The existing color theories represent first approximations to the actual situation. Experiments by Edwin Land, the inventor of the polaroid camera, show that even when only two colors are actually present a great abundance of colors can be perceived and that in some circumstances colors are perceived that are not present at all. Familiar objects, such as a green field, are perceived with astonishing color constancy (by central nervous processes, retinex theory). It is also unclear how metallic colors (gold, silver) are produced; they cannot be mixed from spectral colors

Model of the color system in the retina that encompasses the trichromatic and the opponent process theories (modified after De Valois, 1969 from Birbaumer and Schmidt, 1996)

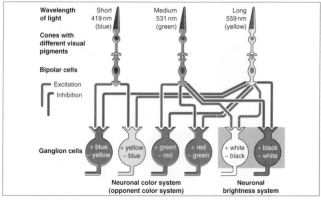

Defects of the color sense, color blindness

Examination: method to test for all color vision defects: pseudoisochromatic charts. They are patterns of many colored dots, arranged in such a way that a person with normal color vision sees a particular number, while one with a defect sees no number or an incorrect one (formed by brightness differences). Anomaloscope: color-mixing device for quantitative measurement of the inability to distinguish red from green. The color impression of a monochromatic (sodium) yellow field is matched in an adjacent field by mixing (lithium) red and (mercury) green. People who see red poorly require more red in the mixture, and so on.

	Inability to distinguish red from green
Protanomaly, protanopia	Confusion of red and green because of inadequacy (protanomaly) or complete absence (protanopia) of red sensitivity. Color spectrum shortened at the long-wavelength end: red blindness: confusion of red with black, dark gray, brown and green
Deuteranomaly, deuteranopia	Confusion of red and green because of inadequacy or absence of green sensitivity. Prot- and deuteranomalies are common defects of color vision: about 8% of all men, 0.4% of all women (inherited as recessive trait carried by X chromosome)

	Inability to distinguish yellow from blue
Tritanomaly, tritanopia	Extremely rare, confusion of yellow and blue. Blue–violet end of the color spectrum shortened, perceived as gray to black
	The various '-opias' are all forms of dichromatopsia, because these people require only two colors to describe all the colors of the color space

	Total color blindness
Achromatopsia	Also (inaccurately) called monochromatopsia. Complete failure of the cone apparatus, hence there is only scotopic black–white vision (see p. 51); normal vision in twilight, by day acuity reduced to 1/10 by central scotoma; also oscillatory eye tremor (nystagmus) because fixation is impossible, photophobia because bright daylight is dazzling

Oculomotor and gaze-directing functions

External eye muscles, innervation, main action

Muscle	Nerve	Main action
External (lateral) rectus	Abducens (VI)	Abductor
Internal (medial) rectus	Oculomotor (III)	Abductor
Superior rectus	Oculomotor (III)	Elevator
Inferior rectus	Oculomotor (III)	Depressor
Inferior oblique	Oculomotor (III)	Outward rotator
Superior oblique	Trochlear (IV)	Inward rotator

The eyes make vergence movements and conjugate movements

Vergence movements	
Convergence	Optic axes are parallel when viewing far-away objects. For fixation of close objects the optic axes must converge. This convergence is closely coupled to contraction of the ciliary muscle for accommodation and to pupillary constriction (p. 50). The combination of these three is called the 'near reflex'
Divergence	The reverse movement of the optic axes when fixation shifts from near to far. Vergence movements have amplitudes as great as 5° and last about 1 s

Conjugate eye movements	
Saccades	Jerky eye movements from one fixation point to another when freely looking around; duration of saccadic movement 10–80 ms, separated by fixation periods of 0.15–2 s. Large saccades are usually accompanied by head movements. The visual acuity during a saccade is low; visual perception is suppressed (saccadic suppression)
Smooth pursuit movements	Occur during fixation of a moving object (advantage: image is kept at the position of greatest acuity, the fovea centralis); often combined with following movements of the head
Rotatory movements	Both eyes move in the same direction in the frontoparallel plane when the head is tilted; also, during marked accommodation convergence is accompanied by slight, symmetrical rotatory movement

The eyes are thus continually in motion. Even during fixation periods, there is a slight oscillation or microtremor (amplitude 1–3 minutes of arc, frequency 20–150 Hz). This seems to be an absolute requirement for vision: an image experimentally 'stabilized' on the retina vanishes after a few seconds (complete adaptation of photoreceptors). Stability of the surroundings despite eye movements is an important example of sensorimotor integration, which is accomplished mainly in the gaze control centers of the brainstem (control center for all eye movements).

Nystagmus

• Definition: Rhythmic eye movements consisting of a periodic alternation between slow following movements and saccades in the opposite direction. Direction of nystagmus is indicated by the direction of the saccadic movement.
• Example 1 – optokinetic nystagmus: When looking out of a moving train, the eye fixates on a point in the landscape as long as possible and then, with a 'restoring' saccade, jumps forward (in the direction of travel) to the next fixation point.
• Example 2 – vestibular nystagmus: When fixation is prevented with Frenzel's spectacles and the subject is rotated on a chair, the horizontal semicircular canals elicit nystagmus in the direction of rotation, and when the chair stops turning this is replaced by postrotatory nystagmus in the other direction. In everyday life this vestibulo-ocular reflex serves to stabilize the eyes during head movements.
• Nystagmogram: electro-oculographic recording (p. 51) of nystagmus

Areas in the brainstem and cortex that control conjugate eye movements

Horizontal gaze center	Reticular formation near abducens nucleus in pons. Activation on one side causes horizontal conjugate eye movement toward the same side, e.g. the right horizontal gaze center causes both eyes to deviate to the right. Output for this is by right abducens and left medial rectus motoneurons; control of the latter is through the medial longitudinal fasciculus (MLF)
Lesion of horizontal gaze center	Destruction of the right horizontal gaze center causes an inability to look conjugately to the right, and there is a tendency for the eyes to look left
Internuclear ophthalmoplegia	Due to interruption of MLF. An attempt to gaze conjugately leads to abduction of one eye (intact output through abducens), but failure of adduction of the other eye because of the interruption of the connection with the oculomotor nucleus through the MLF
Vertical gaze center	Located in midbrain. Controls conjugate vertical eye movements (as well as vergent movements)
Frontal eye field	Located in frontal lobe (area 8). Activation of one frontal eye field causes saccadic conjugate deviation of the eyes toward the opposite side. Output is through the superior colliculus and the brainstem gaze centers
Occipital eye field	Located in MT and MST areas of cortex. Involved in optokinetic nystagmus and in smooth pursuit movements. Output includes several brainstem nuclei and cerebellar connections

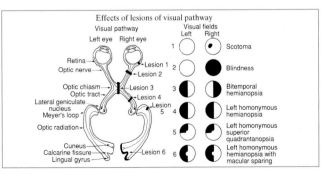

Effects of lesions of visual pathway

Signal processing in the visual system

Signal processing in the retina

Term	Definition/description/comments
Transduction	Photoreceptors are 120 million rods (scotopic vision) and 6 million cones (photopic vision). Both types contain visual pigments in their other segments, rhodopsin ('visual purple') in the rods and 3 pigments in the cones. When the visual pigments absorb light, their conformation changes and they decompose into precursors; by way of a series of intracellular messengers, this event triggers hyperpolarizing receptor potentials
Local retinal neurons	Photoreceptors are the 'entrance' to the retinal neuronal network; ganglion cells are the 'exit', because their axons form the optic nerve. Between them are the horizontal, bipolar and amacrine cells. Signals pass both from photosensors via bipolar cells to ganglion cells and in the perpendicular direction, in horizontal and amacrine cells. All signal processing in these cells is by slow synaptic potentials (no action potentials) except some amacrine cells
Retinal ganglion cells	Ca. 1 million in each eye, hence considerable signal convergence from the 126 million photoreceptors. They generate action potentials that pass to the brain along the optic nerve. Round receptive fields with an inner receptive field center (RFC) and an outer RF periphery with antagonistic connections. Two main types for black–white vision: on-center neurons (discharge increases with light stimulus in RFC, decreases with stimulus in RF periphery) and off-center neurons with the opposite behavior. In light adaptation the center is small and the periphery large, in dark adaptation the reverse, sometimes no periphery at all. Processing of color stimuli occurs in neurons with either red–green antagonism or yellow–blue antagonism, see figure on p. 58

Signal processing in the subcortical visual centers

Term	Definition/description/comments
Visual pathway	Begins with axons of the retinal ganglion cells (see above). From each eye an optic nerve (cranial nerve II) runs to the optic chiasm (decussation) at the base of the skull. Axons from nasal half of each retina cross to the opposite side and, together with ipsilateral axons, form the optic tract, which sends out collaterals to the hypothalamus (suprachiasmatic nucleus), pretectal region and superior colliculus and the lateral geniculate nucleus (thalamic nuclear region). From the LGN the optic radiation passes to the primary visual cortex
Lateral geniculate nucleus (LGN)	Composed of six layers, some associated with the ipsilateral eye (2, 3, 5) and some with the contralateral eye (1, 4, 6). Receptive fields organized as for retinal ganglion cells (see above)
Superior colliculus	Serves for control of reflex gaze movements, especially the saccades. Neurons respond preferentially to moving stimuli, in some cases are direction-specific
Suprachiasmatic nucleus	Controls circadian rhythms. Visual input signals prevailing ambient lighting
Pretectal region	Involved in pupillary light reflex. Also involved in control of gaze direction, especially in vergence movements and smooth pursuit movements

Signal processing by the photoreceptors and neurons of the retina
(according to Grüsser and Grüsser-Cornehls, 1990)

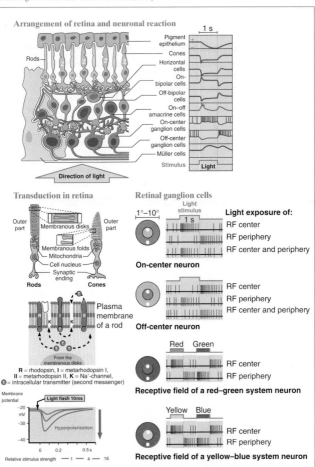

Arrangement of retina and neuronal reaction

Pigment epithelium
Cones
Horizontal cells
On-bipolar cells
Off-bipolar cells
On–off amacrine cells
On-center ganglion cells
Off-center ganglion cells
Müller cells
Rods

Direction of light

Stimulus — Light

1 s

Transduction in retina

Outer part
Membranous disks
Membranous folds
Mitochondria
Cell nucleus
Synaptic ending
Rods   Cones
Outer part

Plasma membrane of a rod

R = rhodopsin, I = metarhodopsin I,
II = metarhodopsin II, K = Na⁺-channel,
= intracellular transmitter (second messenger)

Membrane potential
Light flash 10ms
−20 mV
−30
−40
Hyperpolarization
0   0.2   0.5 s
Relative stimulus strength — 1 — 4 — 16

Retinal ganglion cells

Light stimulus
1°–10°
1 s

Light exposure of:
RF center
RF periphery
RF center and periphery

On-center neuron

RF center
RF periphery
RF center and periphery

Off-center neuron

Red   Green
RF center
RF periphery

Receptive field of a red–green system neuron

Yellow   Blue
RF center
RF periphery

Receptive field of a yellow–blue system neuron

63

Signal processing in the visual cortex

Term	Definition/description/comments
Visual cortex	General term for cortical areas involved in visual signal processing. Optic radiation (see above) terminates in occipital, primary visual cortex (syn.: area 17, area striata or V1). From there information passes toward parietal and temporal, to the extrastriatal areas V2 (area 18), V3, V3a, V4, V5, with increasing specialization for various qualities of vision (e.g. V2 contours, V3 and V5 movement, V4 color)
Retinotopic organization	The whole visual pathway, including the cortices, is topologically organized. Projection is not linear, by far the most space being occupied by the fovea (cf. sens. and mot. homunculus). In higher visual cortices retinotopy decreases in favor of other parameters (see above)
Ocular dominance columns	Information processing in the cortex proceeds perpendicular to the cortical surface (concept of cortical columns, p. 110). In V1 regular alternation between ocular dominance columns processing primarily inputs from the right eye and those processing left eye inputs
Orientation columns	Receptive fields in V1 are oriented lengthwise. The term orientation columns is used for 'subcolumns', within the ocular dominance columns, containing neurons with receptive fields oriented/organized in the same direction. There is a stepwise change in orientation angle from one orientation column to the next
Color columns ('blobs')	Additional 'subcolumns' in the ocular dominance columns, with concentrically organized receptive fields oriented by color (rather than direction)
Forms of the receptive field (RF)	Directionally oriented neurons have either simple RFs with longitudinally arranged on and off zones (respond well to light–dark contours in certain orientations), complex RFs (respond best to interruptions of contour, especially with moving stimuli) or hypercomplex RFs (respond best to moving contrast boundaries that meet one another perpendicularly)

Motion, form and color

Magnocellular system	Contributes to analysis of motion and spatial relationships, as well as to depth perception. Involves large M-type retinal ganglion cells, magnocellular layers of the lateral geniculate nucleus, and several areas of the visual cortex, including V1, V2, V3, V5 (syn.: MT) and MST
Parvocellular interblob system	Contributes to the analysis of form. Involves small P-type retinal ganglion cells, parvocellular layers of the lateral geniculate nucleus, and V1, V2, and the inferior temporal cortex
Parvocellular blob system	Contributes to the analysis of color. Involves P-cells, parvocellular layers of the lateral geniculate nucleus and 'blobs' in V1, as well as V2, V4 and inferior temporal cortex

Peripheral and central vestibular system

The vestibular organ consists of the 3 semicircular canals (with ampullary crests) and the 2 otolith organs (with maculae: saccule and utricle); its function is to detect linear and angular accelerations

Adequate stimulus	Associated with	Part of vestibular organ
Linear acceleration	Positive and negative linear accelerations of all kinds (e.g. gravity, in elevator, in vehicles)	On each side one saccule (normal position vertical) and one utricle (normally nearly horizontal)
Angular acceleration	Angular acceleration about all three spatial axes (e.g. turning the head)	Semicircular canals with their ampullae (one horizontal and two vertical on each side)

The force of gravity (gravitational acceleration) operates continuously; therefore the maculae can always signal precisely the position of the head in space; also any tilting of the head in the earth's gravitational field (these are their most important functions); together with the sensors for proprioception the otolith organs provide the information necessary to determine the position of the body in space. Semicircular canals and otolith organs interact with proprioception to monitor the movements of head and body in space (see p. 36).

Discharge behavior of the hair cells: their spontaneous activity increases when the cilia are bent toward the kinocilium and is inhibited when they bend away from the kinocilium (directional sensitivity of each hair cell)

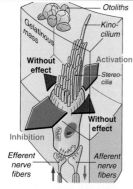

Each hair cell has one kinocilium and 60–80 smaller stereocilia; the cilia are embedded in a gelatinous mass: the cupula in the semicircular canals and the otolithic membrane, which contains otoliths (calcite crystals), in the maculae; the efferent innervation adjusts the sensitivity of the hair cells (functional significance unknown).

Linear acceleration displaces the mass above the maculae, which has higher specific gravity due to the otoliths, and thereby modulates the spontaneous discharge of the hair cells.

Angular acceleration exerts a shear force on the cilia in the ampullae by deflection of the cupula by the inertia of the endolymph; because the three semicircular canals are roughly perpendicular to one another, all conceivable rotational accelerations can be detected.

Because the cilia are oriented in various directions in the array of sensors in the vestibular organ, for every form of acceleration there is a particular constellation of excitation in the associated nerve fibers, which can then be evaluated in the central parts of the vestibular system (see below).

Central components of the vestibular system

The afferent nerve fibers of the vestibular organ have their somata in the vestibular (Scarpa's) ganglion. The axons form the vestibular nerve, which joins the acoustic nerve of the inner ear to form the vestibulocochlear nerve (VIII); this enters the brainstem at the cerebellopontine angle; here the vestibular afferents end in the following vestibular nuclei:

- Superior
- Medial
- Lateral
- Inferior

The vestibular nuclei receive additional inputs from the receptors for proprioception (see above); their main efferent connections are as follows:

- Vestibulospinal tract (primarily to the motoneurons of the extensors)

- To the eye muscle nuclei (by way of the medial longitudinal fasciculus)

- To the contralateral vestibular nuclei (contralateral information exchange)

- To the archicerebellum and other parts of the cerebellum

- To neurons of the reticulospinal tract (from there on to α- and γ-motoneurons, see also p. 95)

- To the thalamus and cerebral cortex (conscious orientation in space).

Function of the vestibular system is to provide information for

1. Conscious perception of the accelerations acting on the body (see above)
2. Maintenance of equilibrium and of an upright stance and gait (especially from the otolith organs)
3. Keeping a fixation point during eye movements
4. Stability of the surroundings during movement of eyes, head and body (especially from the semicircular canal organs, see also oculomotor system, p. 60).

The motor functions of the vestibular system are partly based on labyrinthine reflexes; among them are (see also the definition of postural and righting reflexes, p. 96):

Reflex	Definition/description/comments
Static	These are in part postural and in part righting reflexes; originate in the otolith organs; preserve equilibrium during quiet standing, sitting, lying down; the counter-rotation of the eyes when the head tilts away from vertical is also a static reflex. Postural and righting reflexes also originate in receptors in the neck (tonic neck reflexes); there are also visual righting reflexes
Statokinetic	Originate in otolith organs and semicircular canals; occur during movement and are themselves movements; best-known example is vestibular nystagmus (see p. 60 and below); others are the reflex twisting of a cat's body in free fall and the elevator reaction (adjusts muscle tone to vertical accelerations)

Pathophysiological aspects of the sense of equilibrium

Disorder	Description/comments
Motion sickness	Best-known examples are seasickness and car sickness; caused by greater excitation of the equilibrium organ than one is accustomed to
Failure of labyrinth function	Acute failure of one labyrinth: vertigo toward the healthy side, tendency to fall toward the affected side, nystagmus toward the healthy side. Chronic labyrinth deficit is generally well compensated by the visual system and proprioception; bilateral labyrinth deficits practically never occur in humans
Zero gravity conditions	Only gravitational acceleration is lost, all other linear and angular accelerations being preserved; this constellation of stimuli does not occur on earth; during space flight it may cause motion sickness
Pathological nystagmus	(For definition and examples of physiological nystagmus see p. 60.) Pathological forms involve the vestibular (but also the optic and oculomotor) system, e.g. as spontaneous or as oscillatory nystagmus

Vestibular organ function is tested by nystagmograms (see p. 60); either simultaneous excitation of both labyrinths by rotating the subject (rotatory or postrotatory nystagmus) or unilateral test using caloric nystagmus (elicited by placing cold [30°C] or hot [44°C] water in the external meatus of the ear with the head positioned so that the horizontal canal is vertical).

Psychoacoustics

Acoustic and auditory terminology

- Tone: Sound event comprising only one frequency in the audible range (between 20 and 16 000 Hz); pitch increases as frequency rises
- Harmonic: Tone, the frequency of which is an integer multiple of the fundamental tone
- Complex sounds: Superposition of a relatively small number of tones and their harmonics produces sounds with a musical quality; superposition of many unrelated frequencies produces noise (white noise if the number is practically infinite)
- Effective sound pressure p_x in N/m^2 or Pa, see p. 1; effective value (amplitude, intensity) of periodic air pressure oscillations
- Reference sound pressure: The standard reference pressure has been arbitrarily set at $p_0 = 2 \cdot 10^{-5}$ N/m^2; used to calculate sound pressure level, see below; is near the auditory threshold
- Sound pressure level $L = 20 \cdot \log_{10}(p_x/p_0)$ in dB SPL, see next page
- Sound intensity I (sound power density): Energy flux per unit time through a unit area, e.g. the surface of the tympanum, in W/cm^2; is proportional to the square of the sound pressure, so that when given in dB $I = 10 \cdot \log(I_x/I_0)$
- Auditory threshold: Minimal sound pressure level for perception of a tone
- Loudness level: Expressed in phon, equal to the sound pressure level in dB of a 1000 Hz tone that sounds just as loud as the tone for which loudness level is being determined
- Isophone: Curve including all the tones that sound equally loud.

Measurement and hearing of sound

Term	Definition/description/comments
Sound waves	Very small longitudinal pressure oscillations in the air, which can be detected by the ear; their frequency is given in hertz (Hz) (see p. 1); the sound velocity is ca. 340 m/s (1224 km/h, Mach 1); the audible frequency range is 20–16 000 Hz (<20 Hz infrasound, >16 000 Hz ultrasound); the pitch increases with increasing frequency
Sound pressure level	The dynamic range of the ear encompasses sound intensities from 10^{-16} to 10^{-4} W/cm^2 (12 orders of magnitude from the normal auditory threshold to noises at the pain boundary); hence it would not be practical to use sound pressure or sound intensity as measures; it was therefore agreed to use sound pressure level (SPL), expressed in decibels (dB), definition on p. 67; this gives manageable numerical values between 0 and 120 dB SPL
Loudness	The subjectively experienced loudness of sound is frequency-dependent (see Fig.); for a given sound pressure, sounds between 2000 and 5000 Hz are perceived as louder than higher- or lower-frequency tones; the auditory threshold and loudness are measured by audiometry with pure tones (presented to each ear separately by way of headphones)
Loudness level	Defined on p. 67; it follows that at 1000 Hz the phon and dB values are equal. Curves of equal loudness level are called isophones (see Fig. below). The auditory threshold is also an isophone: the lowest one, at 4 phon. The 60 phon curve passes through the main speech region, the 130 phon curve lies at the pain threshold

The operating range of the human auditory system, represented by curves of equal loudness level (isophones)

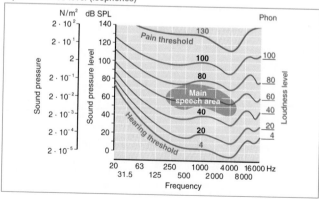

Clinical tests of hearing ability

Clinical threshold audiometry

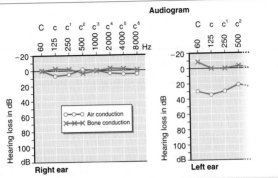

In clinical routine it would be impracticable to represent the complex auditory threshold curve in dB SPL; hence the auditory threshold is arbitrarily assigned the value 0 dB HL (hearing level), making the clinical auditory threshold curve a straight line. By convention, higher threshold values are plotted downward as hearing loss. Conduction in air is tested with headphones, bone conduction by placing an oscillating object on the mastoid process.

Tests of hearing

Test	Description/comments
Weber's test	An oscillating tuning fork is placed on the middle of the skull; in case of inner ear deafness (sensorineural hearing deficit) the tone is perceived as louder on the healthy side, in middle ear deafness (conductive hearing deficit) it is louder on the affected side
Rinne's test	Comparison of air and bone conduction with a tuning fork; it is first placed on the mastoid process; when it can no longer be heard there it is held just outside the ear; the tone is again heard by the healthy and those with sensorineural deficits (Rinne positive) but not by those with conduction deficits (Rinne negative)
Speech audiometry	To test the understanding of speech, prerecorded spoken numbers or standard syllables are played back; in cases of inner ear damage understanding is impaired
ERA	Evoked response audiometry; recording of the auditory evoked potential produced by click stimuli (AEP, see Fig. p. 114): enables objective test without patient's collaboration; especially useful to reveal retrocochlear damage

Functions of the middle and inner ear

Role of the middle ear: impedance matching

Sound is transmitted through the external meatus to the eardrum and from there by way of the hammer (malleus), anvil (incus) and stirrup (stapes) to the perilymph in the scala vestibuli (see Fig.). Without the middle ear only 2% of the sound would enter the inner ear and 98% would be reflected, because air has a lower impedance than liquid; the chain of ossicles transmits 60%, or 30 times as much. This impedance matching is achieved by (1) pressure amplification when force is transmitted from the large eardrum to the small stapedial footplate (area ratio 35:1) and (2) lever action in the ossicle chain.

Sequence of events during sound transmission in the inner ear

Concept	Description/comments
Pressure transfer to the inner ear	The stapedial footplate transmits the sound oscillations to the perilymph of the scala vestibuli, and from there the oscillations propagate through the helicotrema to the scala tympani; because liquids are incompressible, the volume displacements are compensated by movements of the secondary tympanic membrane in the round window
Formation of a traveling wave	The pressure oscillations in the inner ear are also transmitted to the cochlear partition (composed of the organ of Corti, tectorial membrane, scala media, Reissner's membrane; see Fig.), in which they produce waves that travel from the stapes to the helicotrema; the oscillation is always maximal in a particular region, which is located closer to the stapes the higher the tone is
Place principle	(Place theory, tonotopy), states that good frequency discrimination by the ear (e.g. 0.3% at 1000 Hz, hence 3 Hz) is based on the traveling wave maxima in the cochlear partition, because it is only there, at the place associated with the characteristic frequency (CF), that a few hair cells are stimulated
Cochlear amplifier	The passive oscillation maxima induce active contractions of the three rows of outer hair cells (>100-fold amplification); this produces particularly sharp traveling wave maxima, which are the basis of the ear's acute frequency discrimination (see above); loss of the outer hair cells greatly reduces the maxima; the impaired frequency selectivity (flat tuning curves) causes difficulties in understanding speech (the outer hair cells have hardly any afferent innervation and release little transmitter; hence they have predominantly a motor function)
Threshold of the inner hair cells	The threshold of the inner hair cells (only one row) is 50–60 dB higher than that of the outer hair cells; therefore they are not excited until the sharp traveling wave peak has been formed by contraction of the outer hair cells
Adequate stimulus for the hair cells	All hair cells bear stereocilia (sensory hairs that give the hair cells their name); the longest ones on the outer hair cells just touch the gelatinous tectorial membrane; during oscillation of the cochlear partition these cilia are deflected outward and inward by the relative movement between the organ of Corti and the tectorial membrane; this is their adequate stimulus; by a hydrodynamic coupling to this relative movement the inner hair cells are also excited, in a similar manner

Reception, conduction and processing of sound in middle ear and inner ear
(modified after Birbaumer and Schmidt, 1996)

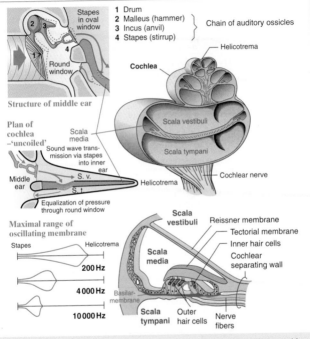

The cochlea resembles a tube coiled through 2.5 turns, like a snail shell; the cochlear partition divides the tube into the upper scala vestibuli (begins at the oval window) and the lower scala tympani (begins at the round window), which communicate with one another at the helicotrema; they are filled with perilymph. The cochlear partition is the actual functional unit of the cochlea. It is bounded by Reissner's membrane on the side of the scala vestibuli and by the basilar membrane on the scala tympani side; the scala media in between contains endolymph (different composition from perilymph, see below); the basilar membrane bears supporting cells and the hair cells, and together these three structures form the organ of Corti; humans have ca. 3500 inner and 12 000 outer hair cells; they are covered by the tectorial membrane; the hair cells are secondary sensory cells, innervated by nerve fibers from the spiral ganglion; there is also an efferent innervation of the hair cells. Each hair cell bears 80–100 sensory hairs (stereocilia, there is no kinocilium).

Prerequisites and mechanisms for transduction in the hair cells

Term	Description/comments
Endocochlear potential (steady potential)	The endolymph of the scala media (1) is extremely K^+-rich (140 mM) and (2) maintains a steady positive potential of +85 mV with respect to its surroundings (e.g. the perilymph), the endo-cochlear potential; (1) and (2) are produced by energy-consuming Na^+,K^+ pumps in the stria vascularis
Resting potential	The resting potential of the inner hair cells is −40 mV, that of the outer ones is −70 mV; hence there are potential differences of −125 and −155 mV, respectively, between hair cell interior and endolymph
Receptor potential	Deflection of the stereocilia opens ion channels at their tips through which, because of the concentration and potential relationships (see 1 and 2 above), mainly K^+ ions flow into the interior and thereby depolarize the cell; this depolarization is the receptor potential; the depolarization in turn opens K channels in the basolateral membrane, through which K^+ ions passively enter the perilymph of the scala tympani; this ends the receptor potential; the whole cycle lasts barely 1 ms
Transmitter release (transformation)	90% of the afferent synapses of the auditory nerve are located at the basal ends of the inner hair cells; here each receptor potential releases glutamate (or a similar substance), which postsynaptically elicits action potentials in the auditory nerve fiber; the rate can be up to 5000 Hz, and this process is strictly coupled to the rhythm of cilia deflection, i.e. to the frequency of the sound signal

Stimulus-elicited potentials of inner ear and auditory nerve (microphonic and compound action potentials) (from Klinke, 1990)

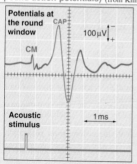

Even in humans, fine electrodes can be inserted through the tympanic membrane and placed on the promontory (bony wall of inner ear) or the round window; then, when sound (e.g. speech) is presented, microphonic potentials (CM, cochlear microphonics) can be recorded, which when played over a loudspeaker reproduce the speech accurately; they originate at or near the outer hair cells; the mechanism is unknown. When the ear is stimulated with a very brief sound pulse ('click'), the compound action potential (CAP) of the auditory nerve can be recorded.

Clinical promontory tests are carried out in order to learn whether the deaf are candidates for an electronic cochlear implant (includes a speech processor and electrodes that can be inserted into the cochlea to stimulate the auditory nerve directly); sometimes these even make it possible to use a telephone.

Otoacoustic emissions

The inner ear produces sounds that can be measured as otoacoustic emissions with a microphone in the external meatus. For example, if a 'click' is used as stimulus, after brief latency a transiently evokable otoacoustic emission (TEOAE) can be recorded; the sound pressure level of the TEOAE is below the auditory threshold; this serves as a screening method to examine hearing in the newborn, infants and small children. Many people also have permanent spontaneous otoacoustic emissions (SOAE). Causes of SOAE and TEOAE are presumably the movements of the outer hair cells (see p. 70), which generate so much energy that some is released to the outside.

Auditory signal processing

Encoding of sound in the auditory nerve fibers (after Klinke, 1990)

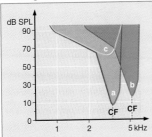

Each nerve fiber coming from the inner hair cells can be optimally excited by sound at a particular frequency, its best frequency or characteristic frequency (CF) (place principle, see p. 70). To excite the fiber with an adjacent frequency, considerably higher sound intensities must be used; this can be represented as tuning curves (threshold curves, **a** and **b**); if the cochlea is damaged, both threshold and frequency selectivity are impaired (**c**).

Periodicity analysis of sound frequency: In addition to the traveling wave the organ of Corti can analyze frequency by detecting periodically recurring sound pressure peaks (periodicity analysis); this must play a role especially above 5000 Hz, because the release of transmitter from the inner hair cells and the action potential bursts of the auditory nerve fibers cannot follow at higher frequencies (see p. 72).

Auditory sensing in three dimensions with binaural hearing (e.g. directional hearing)

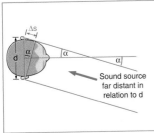

Sound source far distant in relation to d

Differences in arrival time and pressure level arise because one ear is usually further away from the sound source than the other ear; these differences are evaluated for three-dimensional hearing. Arrival time differences as slight as $3 \cdot 10^{-5}$ s can be reliably judged (sound source at $\alpha = 3°$ away from the midline); processing is done by the superior olivary complex. The directional characteristic (distortion of the sound signal) of the pinna is also employed for auditory orientation in space (evaluation mechanism unknown).

Elements of the auditory pathway and their arrangement (after Klinke, 1990)

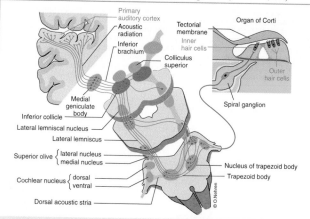

Only the pathways from one ear are drawn here; those from the other have a mirror image arrangement; the descending pathways are omitted, as are the projections to the ipsilateral auditory cortex. Afferent impulses from the organ of Corti are relayed by 5–6 synapses on the way to the primary auditory cortex.

Signal processing in the neurons of the auditory pathway

Term	Description/comments
Sound duration	The length of a sound stimulus is encoded in the duration of activation of the afferent nerve fibers
Sound pressure level	Differences in sound pressure level are reflected in a change in discharge rate of the active neurons; the higher the level, the higher the rate; when the highest possible rate has been reached, further increases in level are signaled by activation of neighboring fibers (recruitment)
Effective sound	The (small) proportion of the sound information that reaches the auditory cortex; for humans the most important effective sound is speech; the background noises, etc. (interference) have already been filtered out at the preceding stations of the auditory pathway, which are specialized for sound pattern recognition
Efferent nerve fibers	The efferent nerve fibers of the auditory nerve mostly originate in the contralateral superior olive (olivocochlear bundle) and 90% of them terminate at the outer hair cells; transmitters to these are acetylcholine and GABA; probably control or modulate the motility of the outer hair cells, e.g. to protect them from sound damage or improved signal detection; details are unknown

Sense of taste

Characteristic properties of the sense of taste

Term	Definition/description/comments
Receptor	Secondary sense cells (taste or gustatory cells, life span ca. 7 d, replaced from basal cells) in the taste buds; in youth humans have ca. 2000 taste buds, each with 40–60 gustatory cells, in old age only $^1/_3$ as many
Locations	The taste buds are situated in the walls of three different types of tongue papillae: the fungiform (over the whole surface of the tongue), foliate (back lateral edges of tongue) and vallate papillae (7–12 in V-shaped arrangement at base of tongue)
Innervation, gustatory pathways	Fungiform papillae: chorda tympani (from VIIth nerve); foliate and vallate papillae: IXth nerve; CNS stations: (1) solitary tract; (2) nucl. of solitary tract; (3) (a) medial lemniscus → ventral thalamus → postcentral gyrus; (b) hypothalamus → amygdala → stria terminalis
Adequate stimulus	Organic and inorganic, usually not volatile substances. Stimulus source near or in direct contact with taste buds
Qualitatively discriminable stimuli	Only four basic qualities (primary taste sensations): sweet, sour, salty, bitter; also mixed qualities, e.g. sweet–sour; under discussion are alkaline and metallic tastes
Topography	Previous assumption: sweet taste at tip of tongue, sour and salty at edges, bitter at base of tongue; new finding: every papilla is sensitive to several taste qualities, usually all four
Sensitivity; adaptation	Absolute sensitivity of taste is low; 10^{16} or more molecules per ml solution needed for suprathreshold stimulation (nevertheless, quinine sulfate tastes bitter at 0.005 g/l). Sense of taste exhibits pronounced adaptation
Biological function	Short-distance sense; to check food for indigestible or poisonous substances; participates in the reflex control of secretion (amount and composition) by the digestive glands

Quality discrimination is based on graded specific discharge of the gustatory nerve fibers, the taste profiles, which are decoded in the CNS (Fig. after Altner, 1985a)

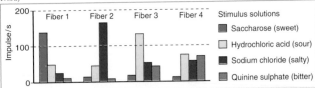

Taste profiles of four chorda tympani fibers (rat) in response to the indicated solutions. The crucial information about taste quality and intensity is thus contained in the afferent impulse pattern of the population of fibers.

Transduction mechanisms in the gustatory cell membranes

Stimulus quality	Transduction mechanism/comments
Sour	Adequate stimulus: H^+ ions; these specifically block K^+ channels in the microvillar membrane of the taste cells; because of the concentration and electrical gradients (see Table p. 11), this causes depolarization, i.e. receptor potential
Salty	Food containing salt causes an increased influx of Na^+ ions through the channel for small cations present in the membrane, which produces a receptor potential. Na^+ is subsequently removed by the N^+,K^+-ATPase pump (see p. 4)
Bitter	There are specific receptor proteins for bitter substances. Their activation initiates a second messenger chain that ultimately raises the intracellular Ca^{2+}. This in turn opens cation channels through which Na^+ flows in to depolarize the cell
Sweet	The various sweet substances all interact with a specific receptor protein, activation of which produces cAMP as a second messenger (for chain of actions see p. 9). The latter blocks K^+ channels and thus produces a receptor potential (see above)

Sense of smell

Characteristic properties of the sense of smell

Term	Definition/description/comments
Receptor	Primary sensory cells, called olfactory cells (total ca. 10 million, life span ca. 30 d, replaced from the basal cells); bear apical cilia that spread out within the mucous membrane
Location	The olfactory cells plus supporting and basal cells together form the olfactory epithelium, which in humans covers ca. 5 cm^2 on the upper concha of each of the two nostrils
Innervation, gustatory pathways	The unmyelinated axons of olfactory cells emerge from the olfactory epithelium as the fila olfactoria; after passing through the lamina cribrosa these form the olfactory nerve (cranial nerve I), which ends in the olfactory bulb. CNS stations: (1) olfactory tract; (2) (a) through the anterior commissure to the contralateral bulb; (b) to the olfactory brain (olfactory tubercle, prepiriform area, amygdala, entorhinal region); from there (3) (a) thalamus → orbitofrontal neocortex; (b) → limbic system → hypothalamus → reticular formation
Adequate stimulus	Molecules of organic compounds in gas form, which do not become dissolved in the liquid phase until they reach the olfactory cells. The stimulus sources are usually a relatively long distance away
Qualitatively distinguishable stimuli	Ca. 10 000 odors are distinguishable; roughly classified into 7 odor classes (see p. 78) each with typical 'representative odors'; odor quality can change with increasing concentration
Sensitivity; adaptation	Absolute sensitivity of olfaction is high; 10^7 or more molecules per ml air needed for suprathreshold stimulation (in animals often only 10^2–10^3). Sense of smell exhibits marked adaptation and cross-adaptation
Biological function	Far and near sense; important role in social communication (key words: scent marks, individual or group identifier, pheromone); strong emotional component

Transduction mechanisms in the olfactory cell membranes (after Hatt, 1993)

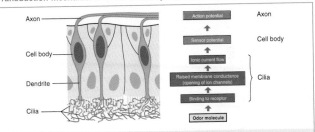

In the membranes of the cilia are a large number of receptor proteins, which upon interacting with the 'appropriate' odor molecule open cation channels by way of the cAMP second messenger mechanism (see p. 9). Activation of a single receptor protein can release thousands of cAMP molecules (hence the low threshold of olfaction). The probability of opening of these channels decreases with increasing intracellular Ca^{2+} concentration. Hence the Ca^{2+} influx after opening closes the channel again: probably a peripheral submechanism of adaptation to odors. A recording of the summed electrical activity of the olfactory mucosa is called an electro-olfactogram (comparable to electroretinogram, etc.).

Information processing in the olfactory bulb (Fig. after Altner, 1985b)

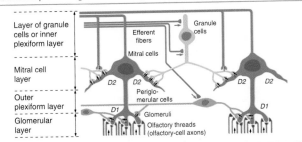

The olfactory bulb comprises four layers. In the glomeruli the olfactory cell axons terminate on the primary (D1) dendrites of the mitral cells (ca. 50 000; their axons form the olfactory tract, see above). The periglomerular cells (also end on D1) and granule cells, which end on the secondary dendrites (D2) of the mitral cells, mediate the efferent inputs and make lateral modulation possible. The direction of synaptic transmission is indicated by arrows (excitation black, inhibition red). The basic features of information processing here are: (1) marked convergence (ca. 1000 olfactory cell axons to one mitral cell), (2) widespread inhibitory mechanisms (transmitter: GABA) and (3) efferent control.

Distinguishing characteristics of the seven odor classes; contributions of nerve V; thresholds for perception and identification; stimulus–sensation relation (Table from Altner and Boeckh, 1990)

Odor class	Representative compounds	Smells like	'Standard'
Floral	Geraniol	Roses	d-l-β-Phenylethylmethylcarbinol
Ethereal	Benzyl acetate	Pears	1,2-Dichloroethane
Musky	Musk	Musk	1,5-Hydroxypentadecanoic acid lactone
Camphorous	Cineole, camphor	Eucalyptus	1,8-Cineole
Putrid	Hydrogen sulfide	Rotten eggs	Dimethyl sulfide
Pungent	Formic acid, acetic acid	Vinegar	Formic acid

Individual olfactory cells have receptors for several odor classes; odor classes, as in taste, depend on a population code. For classification of odors based on congenital partial anosmias see below.

In addition, some substances give rise to olfactory sensations with a stinging, biting or tarry quality (in the nose) or burning sharp sensations (in the mouth). These result from stimulation of free nerve endings of the trigeminal nerve (cranial nerve V), which innervates the entire mucosa of the nose and the oral cavity.

It is typical of olfaction that there is both a perception threshold (unspecific olfactory sensation at very low odor concentrations) and an identification threshold (odor can be identified).

With suprathreshold odor stimuli (the same applies to taste) the intensity of sensation increases with increasing concentration of the odor substance, according to Stevens' power law (see p. 33).

Frequency of occurrence (% of the population) of partial anosmias in man (after Hatt, 1993)

Anosmia for	Main odor component	Frequency
Urine	Androstenone	40%
Malt	Isobutanal	35%
Camphor	1,8-Cineole	33%
Semen	1-Pyrroline	20%
Musk	Pentadecanolide	7%
Fish	Trimethylamine	7%
Meat	Isovalerianic acid	2%

Partial anosmias are astonishingly widespread (see Table).

Affected people seem to lack receptor molecules with which to identify the odors concerned. In view of these clinical observations there are likely to be more than seven, perhaps 10 odor classes.

The partial anosmias are inherited as autosomal recessive traits. Congenital complete anosmia is rare.

When the olfactory threshold is higher than normal the condition is called hyposmia; when it is lower, hyperosmia. Anosmia exists when odor substances are not detected at all. Parosmia designates a qualitatively false olfactory sensation.

Operation of engineering control systems (modified after Schmidt, 1983)

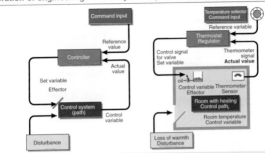

Left: Block diagram of a simple control system, showing the customary terminology (defined below). Right: Illustration of the components of a control system as exemplified in a heating system; for each element of the temperature control system, the corresponding general technical term is also given.

Basic terms in control system technology

- Controlled variable: A state that is to be kept constant (e.g. room temperature, blood pressure, blood glucose level).

- Controlled system: The physical substrate upon which regulation operates (e.g. living room, cardiovascular system, glucose metabolism).

- Sensor: Serves to measure the actual value of the controlled variable (e.g. thermometer, pressosensor, glucosensor).

- Controller: Device that compares the actual value signaled by feedback from the sensor with a reference signal (set point) and, if a difference between them is detected (an error), generates an appropriate control signal (e.g. signal to supply fuel, impulses in autonomic nerves).

- Effector: Apparatus that responds to the control signal so as to correct the error (e.g. furnace with adjustable fuel feed, cardiac output).

- Disturbance: Factor that acts on the controlled system or the controlled variable so as to drive the actual value away from the set point.

Closed-loop and open-loop systems

An essential feature of regulation by a control system is that the circuit is closed, i.e. any deviation from the set point is automatically corrected by negative feedback. It is also possible to employ the same elements in an open-loop configuration, in which there is no negative feedback for automatic error correction. Such a system can compensate for a disturbance known in advance, such as heat loss at a constant outdoor temperature, but not for unpredictable disturbances of variable magnitude.

Operation of biological control systems (from Schmidt, 1983)

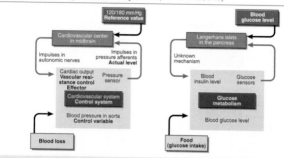

Left: Control system to keep arterial blood pressure constant (see also p. 188); loss of blood is shown as an example of a disturbance. Right: Control system to keep blood sugar level constant (see also p. 144).

Dynamic and static properties of control systems

• Step function: Time course of the response of a control system to a sudden (stepwise) disturbance, its dynamic behavior; some control systems respond rapidly, often actually overshoot and are set into oscillation (e.g. tremor and clonus in the motor system), others are very sluggish; the most important factor here is the gain in the control system.

• Characteristic curves: Static properties of the control system, i.e. relation between input and output quantities in the steady state (e.g. relation between mean blood pressure and mean discharge rate of pressosensors for various static pressure values).

• Regulators and servomechanisms: When the controlled variable is to be kept at a constant reference value, as in the examples in the figures, the system is operating as a regulator, but it is also possible to vary the reference value, e.g. the body temperature during fever; when the reference value changes in this way, the controlled variable must not be affected by disturbance but must follow or 'serve' the new set point, hence it is a servomechanism.

• Feedforward of disturbance variable: Extra signaling of the disturbance directly to the controller, e.g. adding an outdoor thermometer to a thermostat-controlled central heating system, so as to prevent or minimize cooling or overheating pre-emptively; thermoreceptors in the skin operate similarly in the control of body temperature (q.v.).

• Variable-gain control systems: System in which the gain is adjusted to different operating conditions; a large gain keeps departures from the set point small, but can lead to overshooting and oscillations (see above); small gain is more stable, but slower and less accurate.

Molecular mechanism of contraction

Structure and functions of skeletal muscle

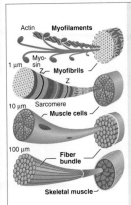

Skeletal musculature makes up 40–50% of the body weight and hence is the largest organ in the body. Main function: contraction, i.e. forceful shortening. Responsible for all communication with the surroundings; also supplies heat.

Skeletal muscle cells (muscle fibers) are very large, multinucleated cells; striated appearance is due to alignment of contractile proteins.

The 'elementary motor' of the muscle is the sarcomere, which is bounded by Z disks and contains mainly the contractile proteins actin, myosin and tropomyosin–troponin, and also other proteins such as myoglobin (for O_2 transport, resembles hemoglobin).

Myofibrils are chains of sarcomeres lined up end to end. Within a muscle cell many myofibrils are bundled in parallel. In turn, the muscle comprises bundles of muscle fibers sheathed in connective tissue.

Sliding filament theory describes elementary processes of contraction

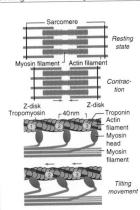

Resting state: Thick myosin filaments in the middle of the sarcomere only slightly overlap thin actin filaments, which project into them from the Z disks.

Contraction: Actin filaments are pulled between the myosin filaments. The force is exerted by cross-bridges (myosin heads) at the ends of the myosin filaments, which attach to the actin, carry out a tilting movement (the actual contraction process, which pulls actin filaments toward the middle of the sarcomere) and then release themselves and 'stretch forward' for a new tilting movement (complete cross-bridge cycle). They then reattach and so on. 50 such rowing strokes produce maximal shortening, bringing the opposed actin filaments so close that they meet in the middle of the sarcomere.

Relaxation: Myosin heads release themselves from the actin filaments, which then slide passively out from between the myosin filaments.

Adenosine triphosphate (ATP) provides energy for the tilting movements of the myosin heads during contraction

For each tilting of a myosin head, one molecule of ATP is split (hydrolyzed) to form adenosine diphosphate (ADP) and organic phosphorus (P). The splitting occurs during the resting state. ATP is needed not for the tilting movement itself (process of contraction, see above), but to release the myosin head from the actin afterward ('relaxing' function of ATP) and extend it for the next tilting movement. The myosin head acts as an ATPase, which is activated by touching the actin filament during the contraction process.

In excitation–contraction coupling Ca^{2+} ions serve as messengers to turn contraction on and off

Structural prerequisite for transmission of the action potential into the fiber is the endoplasmic reticulum. It has two components: (1) the transverse tubules (T system), invaginations from the fiber surface into the interior, and (2) the longitudinal sarcoplasmic reticulum, which expands at each Z disk to form terminal cisternae, in which calcium is stored. Troponin molecules apposed to the actin filaments and tropomyosin filaments spiraling around the actin filaments together interact with Ca^{2+} ions to control access of the myosin heads to actin. Hence the sequence of events in excitation–contraction coupling is:

Event	Description/definition/comments
Propagation of action potential	The end-plate potential transmits excitation from motor axon to muscle fiber (described on p. 19). The resulting muscle action potential propagates at high speed over the whole muscle fiber and penetrates its interior through the T system
Ca^{2+} release	By way of the T system the action potential reaches the sarcoplasmic reticulum, where it releases Ca^{2+} ions into the cell interior (cytoplasm), mainly from the terminal cisternae
Ca^{2+} action	Ca^{2+} ions bind to troponin molecules on the actin filaments; the resulting deformation of these molecules pushes the tropomyosin filaments away from the sites where the myosin heads attach, so that the contraction process can begin
Repolarization of the action potential	The release of Ca^{2+} ions stops, the terminal cisternae pump the free Ca^{2+} ions back again; hence troponin molecules return to the original configuration and tropomyosin filaments again cover the myosin head attachment sites. Tilting movements are no longer possible, and relaxation begins

Before stimulus

-90 mV

10^{-7} mol/l Ca^{2+}
intracellular

20 ms after stimulus

-80 mV

10^{-5} mol/l Ca^{2+}
intracellular

Concentration of Ca^{2+} ions

Action potential

[Ca^{++}]

Stimulus

Contraction

Time [ms]

Muscle mechanics

All forms of contraction can be described in terms of a few basic patterns

The sarcomeres transmit the force they develop through intramuscular elastic structures to the somewhat elastic tendons and to the skeleton. Examples of elastic elements are the myosin cross-bridges and the actin filaments as well as the tendon insertions, which act together as partly parallel and partly serial elastic elements. When the actin filaments are pulled between the myosin filaments during contraction, the serial elastic elements are put under tension and it is this that produces the measurable muscle force.

Form of contraction	Definition/description/comments
Single twitch	Takes about 80 ms to reach maximum, somewhat longer until relaxation; causes only a very small shortening or tension development. Not all muscles twitch equally rapidly. Slow muscles (e.g. postural) contain much red myoglobin ('red' muscles). Fast muscles look 'white' (e.g. eye muscles)
Tetanus (see also p. 85)	During multiple excitation each new single twitch begins while some contraction is still left from the preceding twitch: superposition or summation. At a low excitation rate the single twitches are still discernible (incomplete or partly fused tetanus). At rates above ca. 30 Hz the maximal force is generated (complete or smooth tetanus)

Whether and how much a muscle shortens and how much force it develops depend on the external circumstances under which the work is being done. These too can be described by a few basic patterns; the following characterizations each apply to tetanic contraction; for a comparison of the forms of contraction in heart and skeletal muscle see Fig. p. 169.

Isotonic	Muscle shortening under constant load; there is more rapid shortening, the less the load and vice versa, see Fig. p. 84; the amount of shortening is also greatest with a slight load. The amount of mechanical work done is given by load (force) times distance. If the load exceeds the greatest possible force that can be developed, the active muscle is stretched (typical, very common braking movement, as in walking downhill)
Isometric	Tension developed without muscle shortening (e.g. bracing against a load too heavy to move); maximal force development with prestretching to about the resting length of the sarcomeres (optimal overlap of actin and myosin filaments, see Fig. p. 84). Less force is developed with more overlap or greater prestretching
Ballistic	Shortening with simultaneous force increase (e.g. javelin throwing and other very rapid movements). Rate of shortening is greatest with a small load (see above)
Afterloaded twitch	First isometric, then isotonic contraction (e.g. lifting a bucket); common in everyday life. Again, shortening is more pronounced the less the load. Mechanical work (load times distance lifted) is maximal for an intermediate load (for fatigue-free operation, machinery must be appropriately designed)
Stopped twitch	Initial isotonic shortening followed by isometric contraction after an immovable barrier is encountered (e.g. biting through soft food until the jaws close)

Forms of twitch and the relations between contractile force, sarcomere length and filament overlap (modified from Rüdel, 1993)

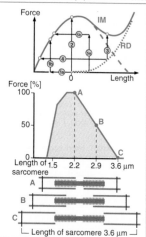

The passive tension curve PT describes the force that must be expended to stretch the muscle passively to a given length. The following forms of contraction can be observed:
1. stopped twitch,
2. and 3. isometric contraction at resting length (2) and with prestretching (3),
4. isotonic contraction,
5. afterloaded twitch.

The curve of isometric maxima IM gives the maximal isometric force achievable by a tetanically excited muscle at a given length. The dashed curve represents IM minus PT, i.e. the actively generated force.

The relationship between contractile force and prestretching depends on the amount of overlap between the actin and myosin filaments. The optimum lies at a sarcomere length of 2.2 µm, because in this case all cross-bridges present are available for attachment and tilting.

The maximal rate of shortening or lengthening of a muscle (red curve) depends on the load that is lifted or is applying tension, respectively (after Rüdel, 1993)

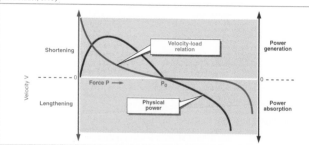

With load 0 the shortening velocity is maximal (isotonic); the maximal isometric force P_0 is developed with a load such that no shortening is possible. Between these limits the velocity–load relation is hyperbolic. With a load > P_0 the muscle lengthens despite contraction (muscle used to 'brake' movement; very common form of muscle activity). The blue curve shows how much physical power a muscle can generate or absorb with different loadings.

Neural control of muscle contraction

Definition of the motor unit (MU); its size

> Every motor axon supplies several to many muscle fibers by way of axon collaterals.
> These elements plus the associated motoneuron are together called a motor unit. Each
> action potential in the motor axon triggers a twitch in all muscle fibers of the MU. The
> smaller the motor unit and the less forceful it is, the more finely graded the contraction
> can be. The size of the MU is extremely variable; for example, the external eye muscles
> have ca. six muscle fibers per MU, while the biceps muscle of the arm has about 750
> fibers per MU.

The gradation of contraction in everyday activity and the development of
muscle tone are brought about by tetanization and recruitment; measured by
electromyogram (EMG)

Event	Definition/comments
Tetanic activation (see also p. 83)	Every MU is capable of gradation from single twitch through partial fusion to complete tetanus. The transition from partly fused to complete tetanus occurs at an excitation rate of $8–30\,s^{-1}$. The force generated rises to tenfold that of the single twitch. Excitation rates between 30 and $120\,s^{-1}$ serve to vary the shortening velocity ($80–120\,s^{-1}$ for only ca. 100 ms at onset of ballistic movements). Our fingers can be moved back and forth up to 8 Hz.
Recruitment	Gradation of the number of activated MUs plays a greater role physiologically than tetanization (see above). An increase in number of activated MUs also raises contraction velocity, because each MU has a smaller load to accelerate (see above). Within a muscle small MUs are activated more often than larger ones (see also p. 21)
Muscle tone	An erect body posture requires continual slight muscle tension without length change (see postural motor systems p. 95). This is achieved by asynchronous activation of MUs. The resulting active background muscle tension is called tone (muscle tone). The degree of tone changes continually, e.g. a distinct decrease in deep sleep and an increase during excitement and mental effort and for heat production (extreme: shivering with cold)
Electromyogram (EMG)	Extracellular recording of the action potentials of single motor units. The electrodes are either attached to the skin over the muscle or inserted into the muscle as needles (electrically insulated except for the tip). Used as diagnostic aid in muscle diseases (myasthenia gravis, myotonia, muscular dystrophy), to measure muscle tone in psychophysiology (especially forehead and upper arm muscles) and in behavioral medicine and rehabilitation to provide feedback of muscle activity for the psychological treatment of tension-related pain and of flaccid and spastic paralysis

Muscle energetics

ATP is the direct source of energy for muscle contraction (see above); efficiency of muscular work; forms of muscle heat

- The reactions following contraction serve to resynthesize ATP (Table). Prolonged muscular activity must be supported by aerobic (i.e. with consumption of oxygen) oxidative phosphorylation. Brief (ca. 30 s) powerful effort can also be supported anaerobically (without oxygen consumption) by way of glycolysis (e.g. sprint at 10 m/s).

- The sarcomeres have a high mechanical efficiency of 40–50%, the remaining energy is dissipated as heat (used to maintain body temperature, see p. 219). The overall efficiency with respect to the exterior is 20–25%, because in addition to the losses during contraction there are also losses in the purely chemical recovery processes. Shivering with cold does no mechanical work and is exclusively for the purpose of heat production.

- The resting heat of resting metabolism is derived entirely from oxidative processes. During work it is supplemented by (1) initial heat during contraction and (2) heat of recovery after contraction; the latter can last for many minutes following powerful muscular effort.

The direct and indirect energy sources in human skeletal muscle
(from Peachey et al., 1983)

Energy source	Content (µmol/g muscle)	Energy-supplying reaction
Adenosine triphosphate (ATP)	5	$ATP \leftrightarrow ADP + P_i$
Creatine phosphate (PC)	11	$PC + ADP \leftrightarrow ATP + creatine$
Glucose units in glycogen	84	Anaerobic: breakdown by way of pyruvate to lactate (glycolysis)
		Aerobic: breakdown by way of pyruvate to CO_2 and H_2O
Triglycerides	10	Oxidation to CO_2 and H_2O

75% of the energy requirement at rest and during stationary work is supplied from fatty acids, the rest from carbohydrates

As the above Table shows, ATP is present in muscle only at low concentration, so that rapid resupply is necessary. The creatine phosphate depot is sufficient for about 100 twitches. During moderate work it is replenished by breakdown of free fatty acids (75%) taken from the blood and by glucose (25%). These proportions reverse only during brief high and maximal performance. The glycogen depots in the muscle cells are drawn upon only for extreme exertion. For further remarks on muscle metabolism and reactions that adjust the body to muscular work, see p. 228.

Fatigue and exhaustion are partly peripheral and partly central in origin

Definitions of the terms fatigue and exhaustion are on p. 227. Depending on circumstances, the fatigue caused by muscular work can be more mental or more physical. There is no fixed borderline between the two; the causes of mental fatigue (fatigue in the CNS) are not yet well understood.

Smooth musculature

Brief characterization of smooth muscle

- Musculature of the viscera and all vessel walls (except for the capillaries, which have no muscle layer).
- The spindle-shaped individual cells are connected by nexuses (gap junctions) to form a syncytium (just like myocardial cells).
- No cross-striation is visible in the light microscope, because the actin and myosin filaments are not arranged in a regular pattern (hence 'smooth' muscle).
- The sliding filament theory (see above) also applies here, but the individual rowing stroke is 100–1000 times slower than in skeletal muscle; this conserves energy for sustained contractions, but makes rapid movement impossible.
- Innervated by the autonomic nervous system. Many smooth muscles are also spontaneously active (myogenic activity).

Important characteristics of smooth muscle; differences from skeletal muscle

Term	Definition/comments
Tetanus	Because the single twitch is very long lasting (often many seconds), complete tetanus is reached at very low excitation rates (<1 Hz)
Myogenic excitation	Pacemaker cells generate spontaneous action potentials, which propagate across the gap junctions and maintain tone without neural input. Acetylcholine raises the spontaneous frequency and norepinephrine lowers it. Hormones also have an influence, e.g. estrogens on the uterine musculature and angiotensin II on vascular musculature; for visceral musculature see pp. 239, 240
Myogenic rhythm	Organ-specific, periodic fluctuations in myogenic tone, responsible for peristaltic waves in the viscera, where it is called basal organ-specific rhythm (BOR) (see p. 236)
Behavior when stretched	Plastic (viscoelastic); when stretched, smooth muscle yields plastically after an initial elastic period. Therefore smooth muscle can be relaxed in both shortened and stretched states (e.g. urinary bladder). In reaction to severe stretching the pacemaker cells increase their activity and elicit contractions (bladder evacuation, autoregulation of arteriole diameter, etc.)
Neurogenic excitation	This predominates in the smooth muscles of arteries, vas deferens, iris, ciliary muscle, which lack pacemaker cells; an inhibitory influence by way of the autonomic nervous system is also common
Excitation–contraction coupling	A T system (see above) is lacking and there is little sarcoplasmic reticulum to store calcium. The cell membrane contains additional action potential-controlled calcium channels. Hence contraction is elicited partly by influx of extracellular Ca^{2+} and partly by release of sarcoplasmic Ca^{2+}. These processes involve regulator proteins (calmodulin, caldesmon, calponin). Much time is required to pump the Ca^{2+} back, so that contraction subsides slowly

Special features of tonic and phasic smooth musculature of the gastrointestinal tract are described on p. 236.

Pathophysiological aspects

Neurogenic muscle disorders include all defects of the motor unit apart from those of the muscle fiber itself. Motoneuron diseases may be degenerative (e.g. amyotrophic lateral sclerosis), inflammatory (e.g. poliomyelitis) or toxic (e.g. tetanus or botulinus toxin). An example of impaired neuromuscular transmission is myasthenia gravis.

The myogenic muscle disorders, in which the defect resides in the muscle fibers, are also called myopathies. They include myotonia, periodic paralysis, progressive muscular dystrophy, metabolic muscle diseases and myositis.

Neurogenic and myogenic muscle diseases together are called neuromuscular diseases. In many cases the underlying pathological mechanisms have been clarified.

Somatosensory inputs to the motor system

Muscle spindles, Golgi tendon organs and cutaneous afferents are the most important receptors for the motor system

Element	Definition/description/comments
	Structure and function of muscle spindles
Intrafusal muscle fibers	These are thinner and shorter than normal (extrafusal) muscle fibers and are enclosed in a spindle-shaped connective tissue sheath; scattered through the muscle in parallel to the extrafusal musculature; especially numerous in muscles for fine movements (e.g. finger muscles); are stretched when the muscle is stretched and relaxed during muscle contraction
Primary sensory endings	Wound around the middle of each intrafusal muscle fiber ('annulospiral' ending). Afferent nerve fiber (one per spindle) is called Ia fiber (see p. 17). Excited by stretching of the central region, hence measure muscle length. Note: intrafusal contraction also stretches central region (see below). Response has dynamic component, and so they also measure rate of change of length
Secondary sensory endings	Wound around intrafusal muscle fiber on either side of the primary ending; their afferent fibers are called group II fibers (see p. 17); they are also stretch-sensitive
Motor innervation	By γ-motoneurons (smaller than α-motoneurons). End-plates of the (thin) γ-motor axons are situated between the poles of the spindles and the afferent innervation. Activation causes this polar region to contract and stretch the central region (thus exciting the primary afferents)
	Structure and function of Golgi tendon organs
Tendon organs	Tendon fascicles comprising ca. 10 extrafusal muscle fibers are enclosed in a connective tissue sheath; present in the tendons of all muscles, arranged 'in series' with the extrafusal musculature and hence are stretched during both stretching and contraction of the muscle
Afferent innervation	By 1–2 Ib nerve fibers (see p. 17), the ends of which branch among the tendon fascicles. They are excited by stretching of the tendon organ and hence measure muscle tension

Cutaneous receptors: Details of the cutaneous receptors and their properties are given on pp. 34–43 and 44–49. Receptors important for escape reflexes (flexor reflex, crossed extensor reflex) are the nociceptors, which are excited by noxious (damaging) stimuli

Discharge patterns of the muscle spindles and tendon organs

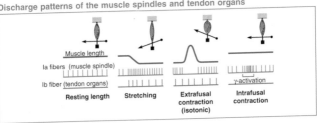

Components and functions of the motor system

Definition of motor centers of the central nervous system (CNS)

The motor nervous system includes all neural structures, the only or predominant function of which is the control of posture and movement. Such structures, e.g. the motor cortex, are called motor centers. These are situated in a cascade arrangement in various parts of the CNS. They are specialized for motor tasks and hence must co-operate with one another. The key words, therefore, are hierarchy and partnership.

Survey of the motor system; explanations of the positions of the individual components and their functions are on the following page (after Birbaumer and Schmidt, 1996)

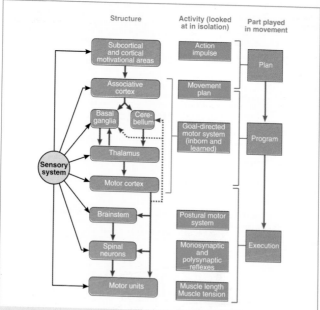

The most important structures and their connections are shown in the left column. For simplicity all sensory inputs are lumped together at the far left. The middle column emphasizes the predominant functions of the individual centers considered in isolation and on the right their roles in initiating and executing a movement are indicated.

Cascade organization of the motor system and the special functions of the motor centers

Component	Position and function(s)/comments
Motor unit	Consists of a motoneuron in the spinal cord or brainstem plus the muscle fibers innervated by its axon; the most peripheral component of the motor system. The size of a motor unit ranges from a few muscle fibers in the external eye muscles to 500–1000 fibers in the back musculature
Anterior horn of spinal cord	Part of the gray matter of the spinal cord; contains the interneurons and motoneurons of many mono- and polysynaptic motor reflex arcs. The spinal reflexes constitute a library of elementary posture and movement programs, which when initiated 'from outside' (sensory stimuli) or 'from above' (voluntary movement) are executed automatically
Motor brainstem centers	Chief among these, from caudal to cranial, are: (1) medullary part of reticular formation (RF), (2) motor components of vestibular nuclei, (3) pontine part of RF, (4) red nucleus. Main role of these centers: (1) control of postural motor functions and (2) co-ordinating these with directed movements, as well as (3) regulating muscle tone
Motor cortex	This includes (1) the primary motor cortex in the precentral gyrus (area 4) and (2) its neighboring motor areas (area 6 with supplementary motor area (SMA) and premotor cortex (PMC)). Part of the pyramidal tract originates here (corticospinal tract). Main role: goal-directed motor programs including (esp. area 4) control of fine movements; area 6 is involved in planning of movement
Motor thalamus	Most important motor nucleus of the thalamus is the ventral lateral nucleus (VL). It links the cerebellum and the basal ganglia to the motor cortex. Main role: incorporating sensory inputs into motor functions
Cerebellum	The main components (with cortical and nuclear regions in each case) and main roles are: (1) vermis: control of postural motor systems; (2) intermediate region: coordination of posture and goal-directed movements; (3) hemispheres: control of rapid (learned, ballistic) directed movements. In addition (1) and (2) participate in oculomotor control
Basal ganglia	Most important components: striatum (input structure, consists of putamen and caudate nucleus), pallidum, and the associated substantia nigra and subthalamic nucleus. Main function: production of motor programs (generation of spatiotemporal impulse patterns to control the amplitude, direction, speed and force of a movement)

The subcortical (e.g. 'hunger center' or 'thirst center' in the hypothalamus) and cortical (association cortex with input into cerebellum and basal ganglia) motivation areas, which are responsible for the internal drive to act and the design of movement strategies, can also be considered motor centers in a broad sense.

To a great extent our motor systems serve to adopt and maintain posture and the orientation of the body in space. In a category apart from these postural functions are the goal-directed movements related to the world outside the body. Goal-directed (voluntary) movements are always accompanied by actions and reactions of the postural motor system.

Motor functions of the spinal cord: reflexes

Every reflex arc consists of the same five elements

Element	Definition, description, comments
1. Receptor	All sensory receptors in the muscles, skin, viscera and the special sense organs (e.g. eye) participate in reflexes of one kind or another
2. Afferent	The afferent nerve fibers of the receptors form the afferent limb of the reflex arc
3. Central neurons	Their number is always >1 except for the monosynaptic stretch reflex (see below). Excitatory and inhibitory inputs to these neurons are the basis of reflex plasticity
4. Efferents	In motor reflexes these are the motor axons; in autonomic reflexes they are the postganglionic fibers of the autonomic nervous system
5. Effectors	In motor reflexes these are the skeletal musculature; in autonomic reflexes, the smooth musculature, the heart or the glands

The monosynaptic stretch reflex (myotatic reflex) is the simplest example of a motor reflex arc (see figure on the following page)

Element	Definition, description, comments
1 & 2	Primary muscle spindle endings and Ia fibers of the homonymous muscle. The reflex arc is activated by stretching the muscle, hence the name stretch reflex
3	α-motoneurons of the homonymous muscle (i.e. the muscle from which the Ia fibers come). The reflex arc has only one central synapse (Ia afferent to motoneuron), hence the name monosynaptic stretch reflex
4	α-motor axons of the homonymous muscle
5	Extrafusal muscle fibers; because these are in the same muscle as the muscle spindles (homonymous) the reflex is also called the myotatic reflex

Has a very short reflex time (time from beginning of stimulus to effector action 20–30 ms). The best known example is the patellar tendon reflex; stretching of the quadriceps femoris muscle by striking the knee tendon. Called an H reflex (after P. Hoffman) when elicited by electrical stimulation of Ia fibers (e.g. tibial nerve behind the knee) recorded myographically as an H wave. Higher-intensity stimuli also elicit an M wave by exciting motor axons. Main reflex function: reflex stabilization of muscle length (postural tone)

General terms for the elements of a reflex arc: specific elements of the monsynaptic stretch reflex

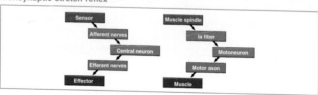

Reciprocal inhibition of motoneurons to antagonist muscles by Ia afferents is the simplex inhibitory reflex arc

Element	Definition/description/comments
1 & 2	Primary muscle spindle endings and Ia fibers of the *antagonist* muscle(s). Activation simultaneous with monosynaptic stretch reflex (on the *agonistic* side)
3	Spinal interneuron with inhibitory synapse on antagonist motoneurons, hence two central synapses (disynaptic reflex arc)
4 & 5	Antagonist motor units are inhibited (decreased tone); also called direct inhibition because of the short pathway. Main function: reinforces monosynaptic stretch reflex in maintaining constant muscle length and joint position

Reflex pathways of the stretch reflex and reciprocal inhibition at a hinge joint involving flexor (F) and extensor (E) motoneurons

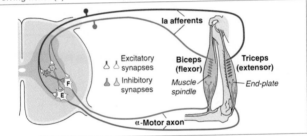

The monosynaptic stretch reflex can also be elicited by intrafusal contraction: γ-loop

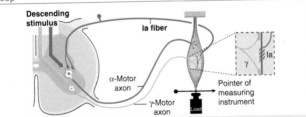

Excitation of the γ-motoneurons by higher motor centers causes intrafusal contraction, which stretches the central region of the muscle spindles and thus activates the monosynaptic reflex arc: γ-loop. Usually α- and γ-motoneurons are excited simultaneously: α–γ-linkage. Advantage: muscle spindle is not taken out of its operating range by excessive shortening or stretching; instead, its sensitivity is adjusted to the new muscle length.

Golgi tendon organs produce effects opposite to those of the primary muscle spindle afferents: autogenic inhibition and antagonistic excitation

Element	Definition/description/comments
1 & 2	Golgi tendon organs and Ib fibers of the homonymous muscle. Reflex arc activated by increased tension (stretching or isometric contraction)
3	Homonymous and agonist motoneurons: there are 1–2 interneurons in the path that inhibit the motoneuron. Antagonist: interneuron with excitatory synapse on motoneuron (di- and trisynaptic reflex arc)
4 & 5	Homonymous and agonist motor units are inhibited: autogenic inhibition. When muscle tension decreases, autogenic inhibition also decreases, causing disinhibition. Antagonist motor units show excitation (increased muscle tone). Main function of both reflex arcs: to maintain constant tension

Flexor reflex and crossed extensor reflex are prototypes of polysynaptic motor reflexes

Element	Definition/description/comments
1 & 2	Nociceptors (high-threshold mechano-, thermo-, chemoreceptors) with group III and group IV afferents (Aδ and C): excited by noxious (i.e. actually or potentially damaging) high-intensity stimuli
3	Spinal interneuron chains (polysynaptic reflex arcs) that excite ipsilateral flexor motoneurons and contralateral extensor motoneurons; inhibit corresponding antagonists
4 & 5	Motor units in ipsilateral flexors are strongly excited, antagonists are inhibited. The result is rapid flexion (withdrawal from the danger zone). Contralaterally the extensors are excited (elevated tone) and flexors inhibited (to strengthen the supporting leg, since the flexed leg no longer provides support)

Many kinds of modification can occur in the interneuronal pathway 3, e.g.: summation of subthreshold stimuli until a reflex is triggered; habituation to repeated stimuli and dishabituation when the stimulus changes; sensitization to painful stimuli, and conditioning, a long-term change in the reflex response due to learning, etc.

Motor functions of the brainstem: postural functions

Motor centers of the brainstem; their efferents and afferents (after Schmidt and Wiesendanger, 1990)

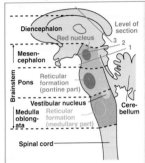

The motor centers of the brainstem control postural functions and muscle tone by way of descending tracts named according to their origin and destination:

1. Lateral reticulospinal tract (from medullary RF).

2. Medial reticulospinal tract (from pontine RF).

3. Vestibulospinal tract.

4. Rubrospinal tract.

A modulating influence is also exerted by the descending noradrenergic system from the locus coeruleus and by the serotonergic system from the raphe nuclei.

These centers receive inputs from the somatosensory system (especially in the neck region), the vestibular system and the motor centers higher in the hierarchy (cerebellum, basal ganglia, motor cortex, see Fig. p. 90)

The functions of the brainstem centers are revealed by eliminating higher motor centers

Situation	Definition/description/comments
Decerebrate animal	Transection at the boundary between midbrain and pons (section plane 1 above); cerebellum intact. The result is a massive increase in muscle tone: decerebrate rigidity, especially affecting the extensors and neck muscles (similar to human apallic syndrome after severe skull/brain trauma or cerebrovascular accident). Marked resistance to stretching of the musculature during movement. The rigidity disappears after transection of the dorsal roots and hence is based on overactivity of the γ-loop.
Midbrain animal	Midbrain also intact (section plane 2 above). The animal can right itself and the rigidity is less pronounced. These functional improvements are due mainly to preservation of the red nucleus
Thalamic animal	Diencephalon also intact (section plane 3 above). The animal makes spontaneous rhythmic stepping movements, though they are automaton-like
Decorticate animal	Basal ganglia also intact. Movement repertoire well preserved though automaton-like; unrestrained, extremely persistent locomotion (animal walks stubbornly into obstacles). Postural and righting reflexes well developed

Motor functions of the brainstem: postural and righting reflexes, postural synergies (long-loop reflexes)

The reduced preparations described on the preceding page have been used extensively to study the function of brainstem centers, and much was learned about reflex adjustments of tone distribution (postural reflexes) and restoration of the normal body orientation (righting reflexes), both of which originate mainly in the proprioceptors of the neck musculature (neck reflexes) and the vestibular system (labyrinthine reflexes). These analyses greatly increased our understanding of motor organization. Today, however, more attention is paid to the incorporation of postural functions into directed movements. In humans the upright gait is of particular interest, as it involves the collaboration of both spinal reflex mechanisms and supraspinal reflex loops (long-loop reflexes) to create postural synergies, which serve in part to prevent directed movements from interfering with stable posture: anticipatory postural synergies.

Functions of the areas of the motor cortex: goal-directed (voluntary) movement

Electrical stimulation of the cortex revealed somatotopy and multiple representation (after Penfield and Rasmussen, 1950)

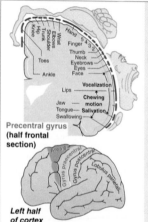

Precentral gyrus (half frontal section)

Left half of cortex

Definition of somatotopy: The body is represented as a distorted map on the motor cortex, especially the precentral gyrus, in which the body regions with a high degree of motor flexibility are disproportionately large; in humans these are mainly the hands and face. Diagrammed as a motor homunculus (for sensory homunculus see p. 42).

Multiple representation: In addition to the precentral gyrus (area 4) there are other somatotopically organized motor cortices, esp. in area 6, supplementary motor area (SMA); premotor cortex (PMC). Distribution of functions among these areas is currently under intensive study. During voluntary movement activity in the SMA (demonstrated by a readiness potential) precedes that in area 4 (demonstrated by a motor potential).

Afferents and efferents to areas of the motor cortex

Inputs come primarily from the basal ganglia (some directly, some by way of the thalamus), the cerebellum (via the thalamus) and the sensory centers (via pontine nuclei). Main efferents: corticospinal tract (pyramidal tract) with collaterals to practically all supraspinal motor centers. Endings partly directly on motoneurons, partly on interneurons of motor reflex arcs, also indirect efferents by way of brainstem centers.

Functional deficits illustrate the roles of the motor cortex

Deficit	Description/comments
Corticospinal tract	(Pyramidal tract) Selective experimental transection causes little impairment of general mobility except for deterioration of fine hand movements: movement slowed, impossible to grip precisely with thumb and index finger, hand grasps by closing all fingers together
PMC	(Premotor cortex) Experimental ablation and clinical lesions cause erroneous postural adaptations; also, movement complexes can no longer be executed in correct temporal sequence
SMA	(Supplementary motor area) Clinical lesion produces distinct poverty of movement and of speech (with normal ability to repeat spoken words)
Cerebral infarct	(Syn.: stroke, apoplexy) Interruption of blood supply in the region of the internal capsule: initially shock with contralateral paralysis (flaccid hemiplegia), later hypertonus (spastic hemiplegia). Many additional symptoms occur (e.g. aphasias, impaired consciousness, spatial neglect). Spasticity resembles decerebrate rigidity. Overall syndrome is caused by interruption of pyramidal and brainstem tracts

Functions of the cerebellum

Afferents and efferents of the cerebellum

1. Medial regions (vermis including flocculus and nodulus, also intermediate region): inputs from the somatosensory, vestibular and visual systems, efferents directly to motor centers in brainstem and spinal cord.

2. Hemispheres: inputs from sensorimotor associative cortex, efferents to motor thalamus (VA) and hence motor cortex (see above).

3. For the cerebellar as a whole: all inputs go to the cerebellar cortex and are doubly represented, as mossy and climbing fibers (functional significance still unclear). The only output from the cerebellar cortex comprises the axons of the Purkinje cells, which make exclusively *inhibitory* synapses (transmitter: GABA) on neurons of the deep cerebellar nuclei (motor modulation by increasing and decreasing inhibition).

Role of the cerebellar structures in motor activities

Structure	Function/comments
Vermis	Control and correction of the postural elements of stance and movement (posture, tone, balance)
Intermediate region	Course correction in slow directed movements and co-ordination of these with the postural elements
Hemispheres	Execution of the rapid directed movements 'designed' by the cerebrum. Assistance in the learning and execution of rapid directed movements without feedback (ballistic movements as in speaking, saccadic eye movement, playing musical instruments, sports)

Functional diagram of the medial parts of the cerebellum (after Schmidt and Wiesendanger, 1990)

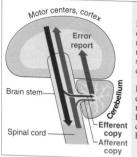

By way of collaterals, these parts receive an efference copy of command signals sent from the motor centers to the spinal cord in descending motor tracts. The cerebellum also receives a sensory afference copy by way of collaterals from ascending tracts.

In the illustrated hypothesis, the cerebellum can calculate departure from the desired values (errors) by comparing the two inputs. The result is fed back to the motor centers to enable continual correction of the motor program once a movement has been initiated.

The motor disturbances and deficits caused by clinical and experimental lesions emphasize the role of the cerebellum

Medial elements
• Flocculus and nodulus: Lesions cause disturbed equilibrium, dizziness, nausea, vomiting, oculomotor impairments such as oscillating nystagmus.
• Vermis and intermediate region: Lesions cause truncal and gait ataxia (as in drunkenness), especially in the dark, when visual control is lacking.

Hemispheres	
1. Asynergy	Inability to send correctly balanced neural signals to muscles. Symptoms as follows:
(a) Movement decomposition	Components of a movement are not executed simultaneously but rather one after another
(b) Dysmetria	Movements are too short (hypometria) or too long (hypermetria) and are subsequently overcompensated
(c) Cerebellar ataxia	Uncertain, overshooting gait with legs wide apart, similar to drunken gait (see above)
(d) Adiadochokinesia	Inability to make movements in rapid succession (as in piano playing) (syn.: dysdiadochokinesia)
2. Tremor	Occurs only during movement: intention tremor; can crescendo in approaching a target so greatly that the target is missed (e.g. in trying to pick up a glass)
3. Hypotonus	Too-low muscle tone, often associated with weakness and rapid fatiguability of musculature

Nystagmus and speech disturbances are also sometimes observed. Charcot's triad: (1) nystagmus, (2) staccato speech, (3) intention tremor. The inability to stop a rapid movement suddenly is called the pathological rebound phenomenon. Lesions in the cerebellar nuclei in general produce more severe symptoms than lesions in the cerebellar cortex.

Functions of the basal ganglia

Afferents and efferents of the basal ganglia

The excitatory afferent inputs come mainly from the entire cortex and sensory centers and run to the striatum (transmitter: glutamate), which sends inhibitory inputs to the substantia nigra and pallidum (transmitter: GABA). Outputs are partly direct to the brainstem, partly to the motor thalamus (inhibitory: GABA) and from there to the motor cortex (as in the cerebellum; there is a marked parallelism of these two structures in the motor system). Internally strong feedback between substantia nigra and striatum (dopaminergic, clinically important: parkinsonism, see below).

Role of the basal ganglia in the motor system

The main function is to participate in the conversion of movement plans into movement programs, i.e. to develop spatiotemporal impulse patterns to control the motor centers that execute movement (see p. 90, right column); includes the setting of movement parameters such as the force, direction, velocity and amplitude of a movement.

Inputs and outputs of the basal ganglia are incorporated in separate, parallel cortico-subcortical, trans-striatal (putamen and caudatum) functional loops, all similarly organized for execution of specific subfunctions

Loop	Function comments
Skeletomotor	Preparation for movement and control of movement parameters (e.g. direction, amplitude, speed, load); somatotopic organization throughout, emphasis on mouth and face movements
Oculomotor	Control of eye movements, e.g. temporal control of saccades; cortical input from the frontal eye field (area 8, supplementary input from area 7)
Complex 'associative'	Three loops known at present: (1) dorsolateral–prefrontal, (2) orbitofrontal, (3) anterior–cingulate. Functions still unclear, probably involved in programs for motor strategies in motivation- and cognition-controlled behavior

Modulation of functional loops in the basal ganglia

The main information flow in the above trans-striatal loops can be enhanced or inhibited by modulation systems.
- Dopamine system: dopaminergic nigrostriatal pathway terminates diffusely in the whole striatum. Discharge rhythm ca. 1 Hz; at each discharge dopamine is released at countless synapses and acts to modulate glutamatergic corticostriatal transmission in the various functional loops (it is unclear, however, whether the action is inhibitory, facilitatory or both).

Other potential modulatory systems:
- serotonergic from raphe nuclei,
- noradrenergic from locus coeruleus,
- cholinergic from striatal interneurons,
- peptidergic from various sources (much still unknown).

Lesions of the basal ganglia cause movement disorders classified clinically as positive and negative symptoms. It may be possible to interpret the two categories pathophysiologically as hyper- and hypoactivity of transmitter systems

	Classification by positive and negative symptoms
Positive symptoms	Hyperkinesias (abnormal, involuntary movements). These include: • athetosis (slow rotating limb movements) • chorea (rapid, brief, irregular movements) • dystonias (slow movements that lead to abnormal postures) • ballism (violent proximal flinging movements); caused by lesion of subthalamic nucleus • tremor (involuntary, strictly rhythmic oscillatory movements of individual parts of the body) • rigidity (abnormal muscle tone during passive movement).
Negative symptoms	Movements disturbed and slowed. Technical terms used essentially synonymously: • akinesia • hypokinesia • bradykinesia (for description of symptoms see below).
	Classification according to transmitter systems
Dopamine	Hypoactivity of the nigrostriatal pathway leads to parkinsonism (see below). Substitution therapy: administration of the precursor L-dopa. Also treated by inhibiting the dopamine-destroying enzyme monoamine oxidase B (MAO-B), and recently by administering dopamine agonists (bromocriptine, lisuride). The disinhibition (hyperactivity) of the cholinergic system produced by dopamine insufficiency is counteracted by anticholinergics (atropine derivatives)
GABA, acetylcholine	Huntington's chorea: hereditary degenerative disease of the basal ganglia (involuntary, tic-like twitches). The GABAergic striatopallidal and -nigral pathways degenerate, as do the cholinergic interneurons (see modulatory systems). So far no substitution therapy is known, nor for hemiballism (unilateral involuntary flinging movements)

Parkinsonism is the most common disease of the basal ganglia. Its symptoms are akinesia, rigor and tremor at rest, to varying degrees

Symptom	Definition/description/comments
Akinesia (synonyms as above)	Difficulty in initiating a movement and in completing it ('freezing' of voluntary movement); face mask-like, expressionless: speech only slightly modulated. No arm movements while walking; small steps with body bent forward
Rigidity	Muscular hypertonia with increased tonic (*not* phasic) stretch reflexes; waxy resistance to passive movement, which periodically gives way ('cogwheel phenomenon')
Tremor at rest	Mainly noticeable in the hands (4–7 Hz), sometimes also lips and other parts of the body; subsides during directed movements and resumes after the movement (cf. intention tremor, p. 98)

Peripheral ANS

Subdivisions of the ANS

Element	Description/comments
Sympathetic	Its preganglionic neurons are all located in the thoracic and upper lumbar spinal cord (red, below); the preganglionic axons (B and C fibers, see p. 17) end either in the paravertebral ganglia of the sympathetic trunk or in unpaired abdominal ganglia; the long postganglionic axons (C fibers) run to the effector organs (see p. 102)
Parasympathetic	Its preganglionic neurons are located in the brainstem and the sacral cord (green below); the preganglionic axons (B and C fibers) end in parasympathetic ganglia near the effector organs, to which the short postganglionic axons (C fibers) run
Enteric	Main elements are the myenteric plexus (Auerbach's) and the submucosal plexus (Meissner's, see p. 236); both contain sensory, motor and interneurons; main function is control of the GIT (pp. 236–249); is independent, but can be modulated by sympathetic and parasympathetic influences

Origin, arrangement and innervation regions of the peripheral ANS (modified from Jänig, 1987)

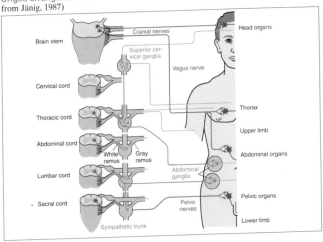

Actions of sympathetic and parasympathetic elements on their effector organs
(after Jänig, 1995)

Organ or organ system	Parasympathetic stimulation	Sympathetic stimulation	Adrenergic receptors
Heart muscle	Decreased heart rate	Increased heart rate	β
		Increased contractile force	β
Blood vessels:			
Arteries in skin and mucosa	0	Vasoconstriction	α
– abdominal region	0	Vasoconstriction	α
– skeletal muscle	0	Vasoconstriction	α
		Vasoconstriction (only by epinephrine)	β
– heart (coronaries)	Vasodilation(?)	Vasodilation (cholinergic)	α
		Vasoconstriction	α
		Vasodilation (only by epinephrine)	β
– genital organs	Vasodilation		
Veins	0	Vasoconstriction	α
Gastrointestinal tract:			
– motility	Increase	Decrease	α and β
– sphincters	Relaxation	Contraction	α
Capsule of spleen	0	Contraction	α
Urinary bladder			
– detrusor vesicae	Contraction	Relaxation	β
– trigonum (internal sphincter)	0	Contraction	α
Internal genital organs	0	Contraction	α
Eye			
– dilator muscle of pupil	0	Contraction (mydriasis)	α
– sphincter muscle of pupil	Contraction (miosis)	0	
– ciliary muscle	Contraction	Slight relaxation	β
Tracheal bronchial musculature	Contraction	Relaxation (mainly by epinephrine)	β
Piloerector muscles	0	Contraction	α
Exocrine glands			
– salivary glands	Much serous secretion	Slight mucous secretion (submandibular gland)	α
– tear glands	Secretion	0	
– digestive glands	Secretion	Decreased secretion or 0	α
– nasopharyngeal glands	Secretion	0	
– bronchial glands	Secretion	?	
– sweat glands	0	Secretion (cholinergic)	
Metabolism			
– liver	0	Glycogenolysis Gluconeogenesis	β
– adipose cells	0	Lipolysis (elevation of free fatty acids in blood)	β
– insulin secretion (from β-cells of islets of Langerhans)	(+)	Decrease	α

Characteristics of ANS operation

- Specific final motor paths: The pre- and postganglionic ANS is organized with functional specificity in its sympathetic and parasympathetic elements; important examples include the systems of the muscular vasoconstrictor neurons, the cutaneous vasoconstrictor neurons, the sudomotor neurons (to the sweat glands) and the pilomotor neurons (to the musculature of the hair follicles).

- Divergence: Within each functional system every preganglionic axon branches to many postganglionic neurons; in this way, a few preganglionic axons can influence many postganglionic axons (distributor and amplifier function).

- Convergence: Within each functional system many preganglionic axons converge onto a single postganglionic neuron; this ensures a high safety factor for synaptic transmission.

- Parallel innervation by sympathetic and parasympathetic systems (see previous page): All organs with parasympathetic innervation also receive sympathetic innervation, but not the reverse (in particular, *no* parasympathetic innervation of the body wall or of vessels).

- Interaction of sympathetic and parasympathetic: In organs with parallel innervation, the two usually have antagonistic actions; however, in some organs the parasympathetic action altogether predominates (e.g. urinary bladder, salivary glands).

- Role of visceral afferents: About 80% of all axons in the vagus nerves and 50% of all axons in the splanchnic nerves are afferent; they innervate the thoracic and abdominal cavities with mechano- and chemoreceptors; their information is used by the ANS together with somatosensory information (e.g. visceroviseral and somatovisceral reflexes).

Synaptic and hormonal transmission in the ANS

Transmitters and receptors in preganglionic and postganglionic synaptic transmission (after Jänig, 1995)

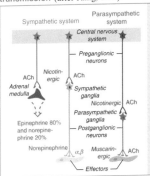

Acetylcholine (ACh) is the transmitter at all preganglionic synapses (nicotinergic receptors, block by quaternary ammonium bases [ganglionic blockers]) and at all postganglionic parasympathetic and some (e.g. sweat glands) sympathetic synapses (muscarinergic receptors, block by atropine, see pp. 5, 7).

Norepinephrine (NE) is the transmitter at most postganglionic sympathetic synapses; postsynaptically partly α-, partly β-receptors (each with diverse selective agonists and blockers), see Table on preceding page and that mentioned above.

Special features of neurotransmission in the ANS; role of epinephrine (E) from the adrenal medulla (AM)

- α-Receptor (syn.: α-adrenoceptor): Pharmacologically defined by the decreasing effectiveness of equimolar doses of the catecholamines NE > E > I (I: isoproterenol, a synthetic catecholamine) and by the effectiveness of specific α-adrenergic blockers.

- β-Receptor (syn.: β-adrenoceptor): Catecholamine effectiveness graded in reverse order, I > E > NE; again there are specific β-adrenergic blockers; the two types of adrenoceptors are currently being classified into subtypes: α_1-, α_2-, β_1- and β_2-receptors, see p. 7.

- Effect of α- and β-receptor activation: Most effectors of the ANS contain both α- and β-receptors (not all are shown on p. 102): their action is usually antagonistic; which action predominates in each case depends on the relative number of adrenoceptors present and on whether they are exposed to more NE or more E (from the AM).

- Release of catecholamines from the AM (see also p. 143): Occurs in the proportions 80% E to 20% NE; slight release in resting conditions, increased under stress, in emergencies and in emotional situations by preganglionic cholinergic activation; in its β-adrenergic action, E serves mainly as a metabolic hormone (glycogenolysis, gluconeogenesis, see p. 102).

- Presynaptic control of transmitter release: Involves feedback by way of presynaptic α- and β-adrenoceptors; partly facilitatory, partly inhibitory; additional reciprocal inhibition between adrenergic and cholinergic postganglionic axons (NE binds to presynaptic α-receptors of cholinergic neurons. ACh binds to presynaptic muscarinic receptors of adrenergic neurons).

- Co-localization of neuropeptides with NE and ACh: Occurs routinely, certain combinations being found more commonly (see p. 6, example in the figure below); classical transmitters and neuropeptides are stored in the vesicles of the varicosities and released from there.

Co-localization of ACh and VIP (vasoactive intestinal peptide) in a postganglionic axon of the salivary gland (modified after Lundberg, 1981. From Birbaumer and Schmidt, 1996)

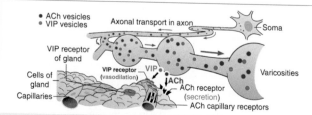

The release of ACh primarily activates the secretion of saliva and to a lesser extent causes vasodilation, whereas in the case of VIP vasodilation predominates.

Spinal and supraspinal organization of the ANS

Arrangement of the spinal autonomic reflex arc (from Jänig, 1995)

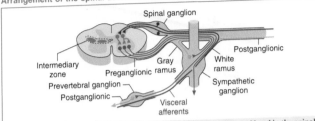

The synaptic linkage between afferents and the ANS at the segmental level in the spinal cord is called the spinal autonomic reflex arc; it comprises at least three synapses, two in the spinal gray matter and one in the autonomic ganglion; there are feedback connections between the receptors of an effector and the effector-specific parts of the ANS, e.g. cardiocardiac and intestinointestinal reflex arcs; somatic afferents can also be the afferent elements for spinal autonomic reflexes (e.g. cutaneovisceral reflex arcs; can be exploited therapeutically, e.g. to improve blood flow through the intestine by applying heat to the skin); conversely, visceral afferents may be connected to motoneurons (viscerosomatic reflex arcs; are responsible, e.g. for increased muscle tone and tenderness associated with intestinal disorders).

Control of the spinal ANS by brainstem and hypothalamus

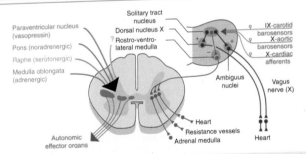

Large parts of the brainstem (medulla oblongata, pons, mesencephalon) and of the hypothalamus participate in the control of autonomic effector organs; the descending pathways that mediate this control and their transmitters are indicated on the left; the right half of the picture shows an example, the structures involved in control of systemic arterial blood pressure (circulatory centers) and their connections; here the sympathetic and parasympathetic elements have functionally synergistic effects (though the individual actions are antagonistic).

Reflex arcs and mechanisms of urinary continence and micturition with pressure–volume diagram (cystometrogram) of the human urinary bladder; example of central nervous regulation by the ANS (modified from de Groat, 1975 and from Simeone and Lampson, 1937)

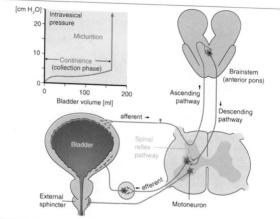

Innervation of the urinary bladder

• Parasympathetic: by way of the pelvic splanchnic nerve with fibers from the 2nd to 4th sacral segments; has excitatory action on the smooth musculature of the bladder wall; prerequisite for normal bladder evacuation.

• Sympathetic: Comes from the upper lumbar cord by way of the inferior mesenteric ganglion (not shown); inhibits bladder wall musculature; hence may contribute to continence.

• Afferent: Stretch receptors in the bladder wall, also nociceptors; afferents run in pelvic splanchnic nerve; are increasingly excited as bladder fills; activate the bladder evacuation reflex.

• Somatomotor: Motor axons of sacral motoneurons, running in the pudendal nerve, supply the external sphincter muscle, i.e. this muscle is under voluntary control.

Continence: Denotes the ability of the bladder to store urine (collecting phase); fills at a rate of ca. 50 ml/h; the urge to urinate begins to be felt when bladder contains ca. 250 ml; can fade away until this rises to ca. 500 ml or be voluntarily suppressed.

Micturition: Denotes the active evacuation phase, preceded by the urge to urinate which results from increased pressure in the bladder (see Fig. above left); once initiated, evacuation of the bladder is completed by way of the pontine reflex pathway; the spinal reflex pathway is normally of no physiological importance (this is not the case, e.g. in paraplegia); voluntary control of micturition is mediated by suprapontine, including cortical structures.

The role of the hypothalamus with respect to the ANS is to organize higher-order autonomic regulation as well as neuroendocrine regulation and elementary behavior patterns; the functional impairments resulting from damage to the hypothalamus in man are correspondingly diverse (modified after Reichlin et al. 1978)

	Anterior hypothalamus, preoptic area	Intermediate hypothalamus (tuberal, ventromedial)	Posterior hypothalamus
Integrative functions	Sleeping–waking rhythm, thermoregulation, endocrine regulation, integration of autonomic and endocrine regulation	Perception, thermal balance, fluid balance, endocrine regulation	Perception, consciousness, thermoregulation, complex endocrine and autonomic regulation
Acute lesions	Insomnia, hyperthermia, diabetes insipidus, ADH secretion disorders	Hyperthermia, diabetes insipidus, endocrine disorders	Excessive sleepiness, emotional disturbances, autonomic disturbances, poikilothermy
Chronic lesions	Insomnia, complex endocrine disorders (e.g. precocious puberty), endocrine disorders resulting from damage to median eminence, hypothermia, no feeling of thirst	Medial: Memory impairments, emotional disturbances, hyperphagia and adiposity, endocrine disorders Lateral: Emotional disturbances, emaciation and loss of appetite, no feeling of thirst	Loss of memory, emotional disturbances, excessive sleepiness, poikilothermy, autonomic disorders, complex endocrine disorders (e.g. precocious puberty)

At many places in this book reference is made to the organ- and system-specific role of the ANS in the context of those structures; here are the associated

Cross-references relating to special aspects of the ANS
(in the sequence in which they are presented in this book)

Definition and localization of integrative functions

For lack of a better term, all functions of the CNS are considered 'integrative' if they are not directly involved in sensory, motor or autonomic activity. There are seven such functions:

(1) Circadian periodicity including the sleep–wake cycle, (2) consciousness, (3) language, (4) thinking (understanding, reason), (5) memory including learning, (6) motivation (drives) and (7) emotion. A crucial structure for (6) and (7) is the limbic system (see next chapter) and for (1)–(5) the cerebral cortex (neocortex or isocortex, usually called simply cortex). Certain cortical areas play special roles, often in only one half of the cerebrum (hemispheric specialization); examples: motor and sensory speech centers (Broca's and Wernicke's areas).

The four lobes of the cerebral cortex (frontal, temporal, parietal and occipital) viewed from the side (from Schmidt, 1990b)

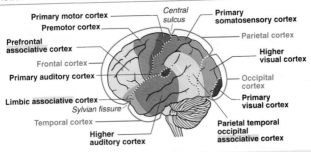

The customary subdivision of the cortex into sensory, motor and association areas. In addition to the primary sensory or motor areas some higher-order areas are shown. Areas are called association or unspecific if no predominant sensory or motor function can be ascribed to them. At present the only areas considered association are those shown here and listed below.

The three large association areas of the human cortex have been found to have the following integrative functions

Association cortex	Integrative function
Parietal–temporal–occipital	Sensory elements of language, higher sensory activities such as linking auditory to visual information
Prefrontal	Higher motor activities such as movement strategies, learned control of innate behavior patterns
Limbic	Motivation, emotional-affective aspects of behavior

General physiology of the cerebral cortex

The figure on the next page shows that the cortex is composed of six layers and contains two main types of neurons: pyramidal cells and stellate cells (from Schmidt, 1990b)

> The pyramidal cells are so called because of the pyramidal shape of the cell body (**A**). Their axons run to other cortical (**B**) or subcortical (**C**, **D**) structures; hence they are cortical efferents. The stellate cells are also named after their shape (**A**). Their axons terminate within the cortex, i.e. they are cortical interneurons. There are many special forms of stellate cells: candelabra cells, basket cells (inhibitory), etc. (**E**).

Differences in the cytoarchitecture (density, arrangement, shape of neurons) of the 6 cortical layers can be represented in brain maps (e.g. in the Brodmann map there are ca. 50 areas)

> These histologically defined cortical fields often coincide amazingly closely with the areas to which particular functions are ascribed on the basis of physiological studies and clinical findings, e.g. the precentral gyrus includes Brodmann area 4, which is the primary motor cortex, and area 6, which includes the SMA and PMC (see p. 96); area 17 is the primary visual cortex, etc. The brain map of Brodmann, first published in 1909, is therefore still in use for orientation to the cerebral cortex.

Layers I–VI (counting from the cortical surface inward) are associated with particular aspects of information processing in the cortex (see also figure on next page)

Layer	Function/comments
I	Contains apical dendrites of the pyramidal cells and tangentially oriented stellate cell axons, which provide local connections between cortical neurons in the immediate vicinity
II and III	Contain small pyramidal cells, the axons of which run to other cortical areas (if these are ipsilateral, the axons are association fibers; if contralateral, commissural fibers) and which receive axons from those areas; subserve intercortical information transfer
IV	Contains stellate cells, the destinations of specific thalamic afferents; subserves thalamocortical information reception
V	Contains large pyramidal cells (especially large ones called giant cells of Betz, found in the primary motor cortex). The axons run as projection fibers to subcortical structures, e.g. basal ganglia, brainstem, spinal cord (from motor cortex: corticospinal [pyramidal] tract); subserves information transfer to subcortical regions
VI	Contains small modified pyramidal cells, the axons of which run to the thalamus as projection fibers; subserves corticothalamic information transmission

> The information processing in the cortex occurs essentially perpendicular to the cortical surface (see neuronal inputs and outputs in **B–D** and their linkage in **E**). This finding has led to the view that the cortex is organized both histologically and functionally into units perpendicular to the surface, the cortical columns.

Simplified schematic representation of the cortical neurons, their circuitry and their afferent and efferent connections, on the background of the layered structure of the cerebral cortex.

A Position and appearance of the two main types of cortical neurons.

B Input/output relations of corticocortical connections.

C Thalamocortical and corticothalamic connections.

D Synaptic input zones of a pyramidal cell.

E Summary of the interconnections of cortical neurons.

B–D drawn from the results of studies by many authors.

Where all six layers are present the cortex is called homotypic; where some layers are not clearly evident it is heterotypical. Heterotypical cortex is called agranular if the small-celled layers II and IV are absent, or granular if II and IV predominate at the expense of the pyramidal cell layers III and V.

The motor cortex is agranular, the sensory areas are granular and the association areas are homotypical.

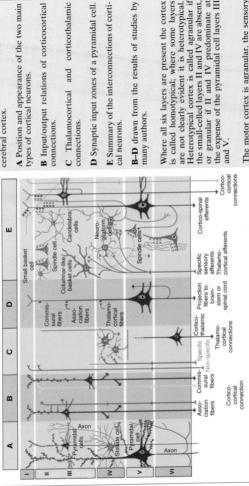

111

The biophysical properties of cortical neurons resemble those of other (e.g. spinal) neurons

Parameter	Definition/comments
Resting potential	−50 to −80 mV, as yet recorded intracellularly mostly in pyramidal cells (also applies to other values in this table)
Action potential (AP)	Amplitude 60–100 mV, duration 0.5–2 ms, generated at axon hillock; no marked afterpotentials, hence AP frequencies up to 100 Hz possible (e.g. in epileptic attacks); secondary generation sites in the dendritic trees, where the APs provide *active* propagation of EPSPs to axon hillock (amplifier function)
EPSP, IPSP	In all cases longer than spinal PSPs: EPSPs 10–30 ms, IPSPs 70–150 ms. The discharge rate of EPSP-evoked APs is low in healthy cortex, usually <10 Hz

When the activity of cortical neurons is recorded extracellularly, with uni- or bipolar electrodes directly on the cortical surface (animal experiment, neurosurgery), the recording is called an electrocorticogram (ECoG); when recorded from the scalp (clinical routine), an electroencephalogram (EEG). Both procedures basically reflect the postsynaptic activity (EPSPs, IPSPs) of cortical neurons; for phenomenology see EEG.

Electroencephalogram (EEG): event-related potentials (ERPs)

Forms of EEG, diagnostic significance

Waves	Definition/comments
	Physiological waveforms
α (alpha)	8–13 Hz; average 10 Hz; basic EEG rhythm at rest, especially with eyes closed, clearest occipitally: synchronized EEG. Rhythmic activity originates in thalamic pacemaker cells; rhythm modified particularly by the reticular formation, e.g. rapid desynchronized EEG while awake (see below) and slow, synchronized EEG during sleep (q.v.).
β (beta)	14–30 Hz, average 20 Hz; rhythm appears when eyes are opened and other sensory stimuli occur, also during emotional excitement (α blockade): EEG becomes desynchronized because frequency is increased, so that its amplitude is reduced
θ (theta)	4–7 Hz, average 6 Hz; in healthy adults appears during sleep (q.v.); in children and adolescents such waves also appear normally
δ (delta)	0.5–3.5 Hz, average 3 Hz; in healthy people only observable during deep sleep (q.v.)
	Pathophysiological waveforms
Seizure waves	Low frequency (2–3 Hz), often with characteristic sequence of pointed and slow waves (spike-and-wave complex); EEG remains the most important method for diagnosis and follow-up of the condition
Flat EEG	Generalized extinction of the EEG, also called isoelectric EEG; is used together with other parameters as criterion for death ('brain death'); often preceded by coma state with very slow, irregular waves

Normal and pathological forms of EEG (modified from Schmidt, 1990b, after recordings from R. Jung)

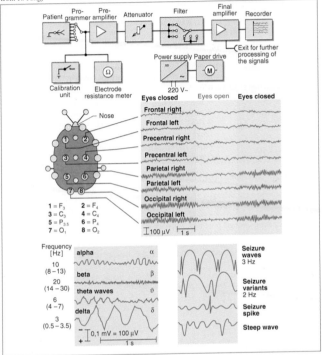

1 = F_3 2 = F_4
3 = C_3 4 = C_4
5 = $P_{3.5}$ 6 = P_4
7 = O_1 8 = O_2

Frequency [Hz]
10 (8–13) alpha α
20 (14–30) beta β
6 (4–7) theta waves ϑ
3 (0.5–3.5) delta δ

Seizure waves 3 Hz
Seizure variants 2 Hz
Seizure spike
Steep wave

Magnetencephalography (MEG); cortical DC potentials

Movement of electrical charge (here: synaptic currents) produces both electrical field changes (recordable as EEG) and magnetic field changes. The latter can be recorded as a magnetencephalogram (MEG); this is an extremely elaborate procedure, because magnetic fields are very small; not practicable for routine use; advantage: better spatial resolution than EEG. Also not routinely applied, for technical reasons: recording of cortical DC potentials (steady potentials); the cortex surface is normally several mV negative with respect to underlying white matter or distant reference electrode; fluctuates when attention shifts and under pathophysiological conditions.

Association between activation and EEG frequency (after Birbaumer and Schmidt, 1991)

Behavioral continuum	EEG frequency	Behavioral efficiency
Very strong activation (emotional excitement)	Desynchronized, low amplitude	Poor; loss of control, disorganized; startle reflex
Moderate activation (attentive wakefulness)	Mixture of high frequencies	Good; effective; selective, rapid reactions
Relaxed wakefulness	Synchronization, distinct α rhythm	Good for automatic reactions (in part for creative thinking)
Dozing, sleepiness	α, θ	Poor, uncoordinated
Light SWS (slow-wave sleep)	θ, sleep spindles, K complexes	Reactions only to very strong stimuli or those corresponding to particular expectations
Deep SWS	δ, large slow waves	
Coma	Large slow waves, also partially isoelectric	
Death	Progressive disappearance of electrical activity; finally completely isoelectric	

Event-related potentials (ERPs) appear in the EEG in association with psychological, motor and sensory events. A special form, **evoked potentials (EPs)** are elicited by sensory stimulation (Fig. from Picton *et al.*, 1974)

EP	Definition/comments
SEP	Somatic evoked potential; appears after stimulation of peripheral somatic nerves or receptors; first positive component, called primary evoked potential, is found only in the topographically associated area of the postcentral gyrus; later, slower secondary evoked potential appears in more extensive cortical regions
VEP	Visual evoked potential; just as complex, highly dependent on nature of stimulus (light flashes, checkerboard pattern, color, etc.): equally important in diagnosis and research
AEP	Auditory evoked potential; complex series of waves that reflect the arrival and synaptic transmission of the afferent burst at the different stations of the auditory pathway; peaks I–VI are called far-field potentials; the N and P peaks originate in thalamus and cortex; AEP is an important diagnostic aid, e.g. where there is a question of hearing loss in infants who cannot yet speak

ERPs, including EPs, are so small in amplitude that they must be extracted from the background noise and other EEG waves by averaging techniques, in which the ERP-evoking event is repeated many times with precise synchronization of the computer processing. Examples of complex ERPs coupled to voluntary directed movement include the expectation potential, readiness potential and premotor positivity.

Imaging methods to reveal the activity, metabolism and circulatory status of the brain

Method	Description/comments
rCBF	Measurement of **regional cerebral blood flow** with radioactive xenon-133; up to 300 detectors distributed over a cortical hemisphere measure gamma radiation, which is higher the more blood is flowing through the cortical area below the detector. The amount of blood flow is directly related to the local energy consumption, and this in turn to the neuronal activity level. The resulting blood flow maps therefore reflect activity of the cortical neurons
CT, CAT scan	**X-ray computer tomography** (CAT scan: computerized axial tomography). Computer-assisted X-ray examination of the brain with a fine beam rotated around the head. Provides high-contrast images of a planar section with a resolution of 0.5–1 mm for a layer 2–13 mm thick. Radiation exposure no greater than in normal X-ray procedure
PET	**Positron emission tomography;** measures release of positrons (or γ-rays, produced when positrons collide with electrons) from radioisotopes that had been incorporated into biologically important molecules (e.g. glucose) and then injected; hence measures the distribution of these isotope-bearing molecules in the brain, which depends on momentary activity (metabolism, in the case of glucose); spatial resolution 4–8 mm, temporal 1 s. Because the isotopes required have short half-lives, the procedure is possible only near a cyclotron; expensive
MRI	**Magnetic resonance imaging;** externally imposed magnetic fields induce resonance of hydrogen atoms (proton) and hence emission of radiation, which is measured and analyzed with computer assistance. The resulting images of brain sections have a layer thickness of 5–10 mm; spatial resolution 1 mm, temporal 10–20 s. These figures are improving with the use of high field strength magnets

Circadian periodicity

Introduction

• Humans (like practically all organisms) have at least two (possibly more) endogenous oscillators (internal clocks): innate rhythm generators that adjust nearly all organ functions to the 24 h day/night rhythm.

• The period of these oscillators is approximately (*circa*) that of a 24 h day (Latin *dies*), hence circadian periodicity. Precise synchronization to 24 h is achieved by external Zeitgeber (entraining signals), especially social factors and the alternation of light and dark. Examples of 24 h fluctuations are found in the body temperature (minimum in the morning, maximum in the evening) and the waking/sleeping rhythm.

• Although certainly innate, in humans a circadian rhythm does not develop until about 15 weeks after birth; after the 'chaotic' early weeks with short, irregular sleeping and waking periods, free-running sleep phases develop which later, from the 20th week on, can be synchronized with the parents' rhythm.

Important properties of the human circadian rhythm

Term	Definition/comments
Circadian	The advantage of a somewhat flexible 24 h periodicity is that it can be adjusted to an altered Zeitgeber (e.g. outdoor temperature) within certain limits, the entrainment range (e.g. the entrainment range for body temperature is 23–27 h, for motor activity 20–32 h)
Free-running	When all Zeitgebers are eliminated (e.g. seclusion from the outside world in a bunker or cave), the circadian periodicity runs with its own rhythm, usually >24 h in humans; the internal clocks mostly remain synchronized with one another, but can also become spontaneously uncoupled (e.g. temperature period remains 25 h, sleeping/waking rhythm shifts to 33 h)
Zeitgeber	This is the German word for environmental factors that synchronize the internal clocks with an exactly 24 h diurnal cycle; in humans these are mainly social factors (knowing the time of day, other people's activities) and the light/dark cycle. A single shift of the Zeitgeber (as in flying east or west) causes a transient disturbance (jet lag); about 24 h of adaptation are necessary per time zone. Night- and shiftwork involve a permanent conflict between personal and ambient Zeitgeber
Localization	Internal clocks are located in the ventral hypothalamus, especially the suprachiasmatic nucleus (SCN) and the ventromedial nucleus (VMN), and perhaps also other sites as yet unknown

In addition to the circadian rhythm there are many others, with periods distinctly shorter or longer than 24 h: ultradian (e.g. respiration, heartbeat, EEG) and infradian, respectively. Chief among the infradian rhythms (e.g. animal hibernation, bird migration) in humans is the female menstrual cycle.

Sleep and dreaming

Four temporal levels for the description of normal and pathological sleep phenomena (modified after Schulz *et al.*, 1991)

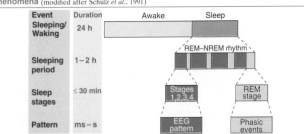

The highest temporal integration level is the circadian sleeping–waking rhythm; the next integration level comprises the sleep cycles with their NREM–REM rhythm; there follow the levels of sleep stages and finally of EEG patterns and phasic events (for details see below).

Time course of sleep cycles in humans in the course of a night (from Jovanović, 1971)

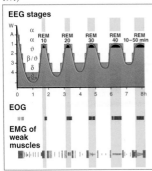

EEG stages

EOG

EMG of weak muscles

The basic structure of sleep is an ultradian rhythm of sleep cycles, each consisting of four NREM stages and one REM stage; average cycle duration 100 min, 3–5 cycles per night.

The five sleep stages are distinguished from one another by the EEG. In each cycle the depth of sleep first increases (NREM stages 1–4) and then decreases.

Each cycle ends with a fifth stage in which there are rapid eye movements (REM stage, recorded by electro-oculogram, EOG); dreams are most likely to occur during the REM stages.

According to the reciprocal interaction model of ultradian sleep periods, aminergic neurons in the brainstem are responsible for NREM sleep and cholinergic neurons for REM sleep (modified after Hobson *et al.*, 1986)

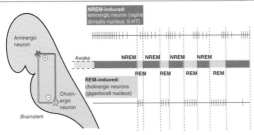

It is not yet completely clear how the transition from waking to sleep and back to waking occurs. Various theories (deafferentation theory, reticular theory, serotonergic theory) are under discussion. As the figure shows, the NREM–REM rhythm is apparently generated by regions of the brainstem; here, again, there is no completely satisfactory interpretation of the many findings.

Endogenous sleep factors are substances that initiate or maintain sleep: factor S is a small glucopeptide, the concentration of which in the cerebrospinal fluid rises during waking and the injection of which induces NREM sleep. DSIP (delta sleep inducing peptide) is a nonapeptide that elicits deep sleep (SS 4).

Characterization of the individual sleep stages (SS; stages 1–4 are grouped together as NREM stages)

SS	EEG pattern/comments
W	Relaxed waking with predominant α rhythm; transition to sleep is sometimes additionally designated stage A
1	Absence of α waves, low rapid β activity and low θ activity; stage of falling asleep and lightest sleep
2	Low rapid activity with β spindles (inhibition of sensorimotor areas), later K complexes (internal discharge of sensory systems); light sleep; more than 50% of sleep is spent in this stage
3	δ waves for 10–50% of the time; intermediate sleep. The sleep cycles in the morning often no longer reach this stage and the stage of deep sleep (SS 4); then sleep passes directly from SS 2 to the REM stage
4	δ waves (>100 µV, <3 Hz) over 50% of the time; deep sleep; together with SS 3: synchronized sleep or SW sleep (slow-wave sleep); musculature atonic, high awakening threshold
REM	Low-amplitude EEG with low θ waves; otherwise the EEG resembles that for the attentive waking state (no α waves), hence the name desynchronized sleep; accompanied by REM bursts (see Fig.), remaining musculature practically atonic, awakening threshold very high. Because of similarity to the waking EEG is called paradoxical sleep (NREM = orthodox). REM duration in first cycles 5–10 min, later longer, in final cycle as long as 22 min (followed by awakening)

Effect of aging on NREM and REM sleep: total sleep time falls throughout life ('senile insomnia'); in addition, proportion of REM sleep becomes considerably smaller (from Roffwarg *et al.*, 1966)

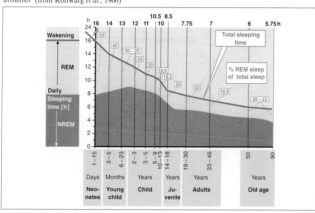

Dreams usually occur during REM sleep, but the brain is mentally active even in NREM sleep; key concepts:

- In 80–90% of awakenings from REM sleep, dreams are reported: the content is substantial, with pictures, odors, tones.

- In ca. 70% of awakenings from NREM sleep, dreams are reported: however, the content is more 'thought-like' (cognitive) than that of REM dreams.

- The REM dreams of the first half of the night are more reality-related and refer to experiences of the preceding day, whereas morning dreams are progressively more bizarre and emotionally intense.

- People remember a dream only if they are awakened within 5 min after a REM period or if it is the last dream before awakening; therefore morning dreams are especially well remembered, which is why the dream life seems so unreal to us

- The psychobiological significance of dreams is still unclear; selective 'dream deprivation' (awakening at the beginning of every REM phase) merely leads mentally to increased irritability and physiologically to prolonged REM periods during the recovery nights

Sleep disturbances that do not result from organic, in particular neurological, diseases are called primary; some common disturbances are summarized below

Name	Definition/comments
	Insomnia (difficulty going to sleep and staying asleep)
Pseudoinsomnia	Subjective difficulties in falling and staying asleep are reported, but the sleep profile is normal for the person's age; usually the disturbance is psychological, e.g. partnership and sexual problems
Idiopathic insomnia	Subjectively experienced and objectively verifiable abnormal sleep profile; many causes, e.g. too much or too little physical activity, chronic stress, traveling, extreme dieting
	Hypersomnia
Narcolepsy	Increased tiredness by day with frequent sleep attacks lasting a few seconds to 30 min; can be interpreted as 'invasion' of REM episodes into the waking state; the attacks are associated with muscular relaxation and loss of tone (sometimes collapse) (cataplexy). There may also be sleep paralysis (loss of muscle tone while sleeping) and hypnagogic hallucinations at sleep onset
Disturbances associated with sleep stages	
Somnambulism	Sleepwalking; motor automatism at the transition from SS 4 to SS 2; seen especially in children and adolescents, also adults under stress. Eyes are wide open, the person does not respond when spoken to and is disoriented after waking up, does not remember dreams
Nocturnal enuresis	Bed-wetting occurs in about 5% of all children after the second year of life; almost always occurs in NREM sleep. After waking up the child is disoriented, confused and cannot report anything about dreams
Night terrors	Occurs mainly between the third and eighth year of life; child sits up and begins to scream; stares with eyes wide open; face pale and covered with sweat; breathing is difficult; similar to nightmares in adults

Consciousness, language, cerebral asymmetry

Consciousness can be (1) experienced introspectively (self-consciousness) and (2) observed in the behavior of others; key concepts:

- There are various forms of consciousness; therefore a unitary definition of consciousness is impossible.
- A physiological characteristic common to all forms of consciousness is an extensive subcortical and cortical increase in excitation (ARAS concept, see below).
- Common psychological characteristic: transition from unconscious 'automatic' to attentive 'controlled' information processing.
- The production of consciousness is a property of short-term memory: processes in long-term memory are not consciously apprehended until transferred to short-term memory.
- Consciousness requires, among other things, activation of the cerebral cortex by the reticular formation (RF); the tracts ascending from the RF are therefore called the ascending reticular activating system, ARAS.

Disturbances of consciousness can be ascribed to three mechanisms, according to the ARAS concept

1. Reversible inhibition or irreversible damage to the RF, as a result of which the central driving force for the cerebral cortex is eliminated.
2. Damage to the ARAS above the RF, with preservation of the functional and structural integrity of the RF itself.
3. Destruction or functional inhibition of the cerebral cortex.

The ARAS concept goes back to observations of sleeping cats, in which high-frequency electrical stimulation of the RF triggered an immediate awakening reaction (arousal), whereas lesions of the RF caused coma-like permanent sleep

Conscious perceptions require simultaneous excitation of the cortical fields (1) by the ARAS, originating in the RF and (2) through the specific sensory pathways (after Hassler, 1979)

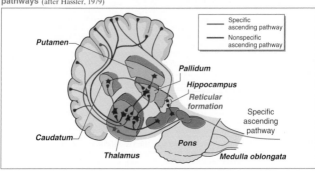

Pathological restrictions of the degree of consciousness can be quantified; various scales are available (e.g. Glasgow Coma Scale, Munich Coma Scale); examples of restricted consciousness:

Name	Definition/comments
Somnolence	Pathological sleepiness; misinterpretation of sensory information, especially visual; thinking (cognition) disturbed and slowed; sometimes temporospatial disorientation, defective memory
Stupor	Impairment of consciousness resembling deep sleep; affected person can be awakened only by strong, repeated stimuli; then falls back into stupor
Coma	Complete absense of response with closed eyes; cannot be awakened even by the strongest stimuli; sometimes diffuse defensive movements in response to painful stimuli
Locked-in syndrome	Complete paralysis (tetraplegia) without loss of consciousness: patient is 'locked' inside his body (though may be able to communicate by eye movements and blinking; result of selective interruption of motor efferents for voluntary movement; can also occur when anesthesia wears off sooner than curare-induced muscle relaxation
Anesthesia	Pharmacologically induced special form of unconsciousness; the mechanism of action of anesthetics is not yet altogether clear; sensory information from the surroundings and the interior of the body cannot be perceived and remembered

In addition to quantitative restrictions of the degree of consciousness, the content of consciousness may be qualitatively affected, as in hallucinations and dazed and confused states; occurs in psychological/psychiatric syndromes (e.g. schizophrenia), in metabolic encephalopathies (e.g. hepatic coma), in epilepsy (aura) and after intake of hallucinogenic drugs, e.g. LSD.

Split-brain operations (transection of the corpus callosum and anterior commissure) produce split consciousness; this is the most spectacular manifestation of cerebral asymmetry; important results:

- Verbal and complex cognitive tasks as a rule can be performed only by the left hemisphere, which also functions as an 'interpreter of causes', i.e. it is responsible for finding non-contradictory explanations for motor and emotional–autonomic reactions (causal attribution, reduction of cognitive dissonance).

- The right hemisphere processes mainly visual–spatial, non-language-related material; it can provide information about this in non-verbal form (e.g. gesturing, pointing, emotional reactions) and hence undoubtedly has a 'consciousness'.

- However, the right hemisphere's form of consciousness is difficult to describe in words; the subjectively experienced unity of consciousness is thus ascribable to the existence of association fibers and other extensive connections in the CNS and the continuous information exchange between the hemispheres that such connections make possible.

- Left hemisphere and speech-related consciousness appear dominant to us, because they allow communication with other people; more precisely, then, we have language dominance on the left, usually associated with right-handedness.

For both production and understanding of speech, several information-processing events in the left hemisphere must take place in sequence (serial) and simultaneously (parallel); the main speech disturbances due to damage of the participating brain structures are:

Speech disorder	Definition/description/comments
Broca's aphasia	Motor (expressive) failure of speech in lesions of the left inferior frontal gyrus (Broca's area); patient speaks very little, only 'key words' (telegraphic style) and only with great effort; articulation and prosody (speech melody) are poor; good understanding of speech
Wernicke's aphasia	Sensory (receptive) failure of speech in lesions of the temporal lobe (Wernicke's area); patient speaks fluently with many phonematic (sound-related, e.g. spilling instead of spinning) and semantic (e.g. mother instead of wife) paraphasias, also neologisms; severely disturbed understanding of speech
Conduction aphasia	Interruption of the arcuate fasciculus, which connects Wernicke's and Broca's areas; verbal communication is possible, but the patient cannot repeat words spoken to him
Global aphasia	Severe disturbance of both expressive and receptive language ability with simultaneous lesion of Wernicke's and Broca's areas
Anomic aphasia	Also called amnesic or semantic; impaired word-finding ability, though speech is fluent and understanding preserved; usually lesions in temporal–parietal cortex
Subcortical aphasia	Initial mutism (dumbness) followed by paraphasias that disappear when repeating spoken words; slight speech production, good understanding of speech, usually rapid recovery
Alexia, agraphia, acalculia	Disturbances of reading, writing and calculating, respectively, that accompany aphasias and can occasionally be the dominant symptoms

A classification of the brain areas involved in language is derived from Geschwind's model; it is based chiefly on evaluation of speech impairments due to brain lesions (subcortical connections are omitted) (based on Kolb and Winshaw, 1985)

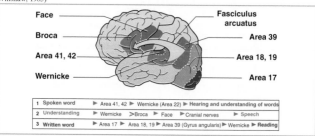

1 Spoken word	► Area 41, 42	► Wernicke (Area 22)	► Hearing and understanding of words		
2 Understanding	► Wernicke	>Broca	► Face	► Cranial nerves	► Speech
3 Written word	► Area 17	► Area 18, 19	► Area 39 (Gyrus angularis)	► Wernicke	► Reading

Plasticity, learning, memory

Learning and retaining behavioral changes in humans and animals
(storage in behavioral memory)

Learning process	Definition/comments
Non-associative learning	
In non-associative learning processes behavior changes as a consequence of repetition of the stimulus situation, not as a result of close temporal association of stimuli and responses	
Habituation	If the internal comparison of an actual stimulus with the 'expected' model of the stimulus reveals a discrepancy, an orienting reaction (OR) (redirected attention, arousal, etc.) results; its intensity is proportional to the mismatch between stimulus and model; habituation is a reduction of the OR when the stimulus is repeated; in humans it is not stimulus intensity but subjective significance that determines OR and time course of habituation
Sensitization	The mirror image of habituation, i.e. an increase of a physiological response or a behavior due to especially intense, in particular noxious, stimulation (high negative significance)
Associative learning	
Learning by conditioning is so called because the central nervous process consists in establishing an association between stimuli (S) and responses (R). The result is a change in behavior (memory).	
Classical conditioning	Acquisition of conditioned reflexes by Pavlov's method: prior to an unconditioned stimulus (US, e.g. food presentation), which elicits an unconditioned reflex (UR, here salivation), a conditioned stimulus (CS, e.g. a bell) is given repeatedly; eventually presentation of CS alone elicits the reflex response (conditioned reflex, CR). Hence CS → US → UR becomes CS → CR. For optimal learning the time interval between CS and US must be <1 s
Operant conditioning	If a desired behavior is (accidentally) exhibited (e.g. bar-press in a Skinner box) and is rewarded within 1 s (e.g. with food), the reward increases the probability that the behavior will be performed again. The behavior itself acts (is operant) to produce the rewarding stimulus; in other words, the behavior is the instrument for obtaining the stimulus (reward): instrumental learning
Imprinting	Special form of associative learning, namely learning of social bonds by a specific stimulus pattern during a temporally limited development period; is manifest in staying close to the 'imprinted' object (mother animal) and in defensive or escape behavior with respect to strange objects; common in birds (geese of Konrad Lorenz); rare in mammals

In humans the acquisition of knowledge (simplest experiment: learning a sequence of nonsense syllables) is called cognitive learning and ascribed to a cognitive memory

In cognitive learning a distinction is made between

- short-term memory (STM), or primary memory (often called working memory) and
- long-term memory (LTM), or secondary memory.
- The transfer of information from STM to LTM is called consolidation; is made easier by practice.
- A memory trace stored in the LTM is called an engram.

The following Tables employ these concepts and supplement them with a

- sensory memory, which precedes the STM, and a
- tertiary memory for especially well-consolidated material.

Survey of the main characteristics of human memory processes in cognitive learning (knowledge memory) (modified after Ervin and Anders, 1970)

	Sensory memory	Primary memory (STM)	Secondary memory (LTM)	Tertiary memory
Capacity	Limited by the information transmitted from receptor	Small	Very large	Very large
Duration	Fractions of a second	Several seconds	Several minutes to several years	Permanent
Entry into storage	Automatic during perception	Verbalization	Practice	Very frequent practice
Organization	Representation of the physical stimulus	Temporal ordering	Semantic and by spatiotemporal relations (Gestalt learning)	?
Access to storage	Limited only by speed of readout	Very rapid access	Slow access	Very rapid access
Types of information	Sensory	Verbal (among others?)	All forms	All forms
Types of forgetting	Fading and extinction	New information replaces old	Interference, proactive and retroactive	Possibly no forgetting

The subdivisions of cognitive memory

Memory	Comments
Sensory	In auditory context called echoic, in visual iconic; serves to increase attentiveness and for feature extraction; encoding for transmission to primary memory is chiefly verbal; forgetting due to passive extinction or active overwriting (erasure) with new information
Primary	(STM) Small capacity but can be increased by organizing the information in chunks (superordinate groups, e.g. letters into words); transmission to secondary memory by practice (memorizing)
Secondary	(LTM) Large, long-term storage-system. Procedural LTM: the learned knowledge itself (e.g. playing a piano from a score). Declarative LTM: knowing that one is able to play from a score and when and how the ability (or knowledge) was acquired. Forgetting due to disruption of the material to be learned by material learned previously (proactive inhibition) or subsequently (retroactive inhibition)
Tertiary	Independent part of the LTM, in which practically unforgettable knowledge is stored, e.g. one's own name and other personal data, the ability to read and write; in contrast to secondary memory, very rapid access to stored material

Neuropsychological disturbances of consolidation: amnesias

Form	Definition, comments
Anterograde	New material cannot be learned, retained and retrieved; primary memory (STM) is usually intact, but transfer to secondary memory (LTM), i.e. the consolidation process, is defective; secondary and tertiary memory normal for the time before onset of the disease; bilateral damage to the hippocampus seems to be the main cause of the lack of consolidation
Retrograde	Loss of ability to remember the time before impairment of brain function, e.g. by concussion, stroke, electroshock; content of primary memory is extinguished, that of secondary memory is lost further back into the past, the more severe the brain damage was; the forgotten time period can shrink during the recovery phase, so apparently access is disturbed more than content; tertiary memory remains intact; the pathophysiological mechanism of retrograde amnesia is unknown
Hysterical	Very rare complete loss of memory, including tertiary memory (one's own name, etc.); exclusively functional, mental disturbance; key stimuli (previous surroundings, family members, friends) have no effect, good STM and LTM for new information
Korsakoff's syndrome	Memory defect in alcoholics, with anterograde and retrograde amnesia; loss of memory is concealed with invented stories: confabulation; no realization of illness, apathy

Neuronal and neurochemical mechanisms of the engrams (memory contents) in the various forms of memory

Memory	Mechanism/comments
Habituation and sensitization	Engram largely laid down as synaptic depression or facilitation (tetanic or post-tetanic depression or potentiation, p. 25). Short-term: reduced or enhanced presynaptic transmitter release. Long-term: structural changes, e.g. increase in size and number of presynaptic active zones
Behavioral memory	Mixture of the above-mentioned synaptic processes; temporary storage by circulating excitation in a 'reverberatory circuit', as a dynamic engram, and eventually cellular consolidation to a structural engram; details largely unknown
Cognitive memory	In STM, a dynamic engram in the form of reverberating excitation, which in the consolidation phase is transferred into a structural engram in the LTM; so far known to involve synaptic potentiation (see above), increase in number of dendritic processes (spines; raises synaptic efficiency), increased protein synthesis (pharmacological inhibition of protein synthesis during critical consolidation phase inhibits formation of LTM engram)

Contrary to all claims, direct and specific pharmacological improvement of intelligence and of the learning and memory functions is impossible; proposed but ineffective substances include glutamic acid (glutamate), cholinergic and anticholinergic substances, strychnine, picrotoxin, tetrazole, caffeine, ribonucleic acid, etc.

Basic concepts related to motivation

Definition of the term 'motivation'

The term 'motivation' refers to factors affecting the frequency and intensity of a behavior that depend on states within the organism, i.e. the behavior observed does not depend exclusively on stimulus, site of stimulation, or genetic predisposition.

Examples of variable states within the organism include departures from homeostatic equilibria (e.g. a lowered blood glucose level, water deficiency) or fluctuating hormone concentrations (e.g. sexual hormones).

Motivated behavior takes the form of drives, distinct behavioral patterns (e.g. foraging and feeding, sexual behavior); the analysis of drives has led to the following concepts

Term	Definition/description/comments
Homeostatic drives	Drives, the intensity of which depend mainly on the degree of deviation of a homeostatic parameter within the body (e.g. blood glucose concentration) from a stable reference point; deviation upward or downward initiates a sequence of behaviors that are performed until the reference point is regained
Non-homeostatic drives	Drives with no clear internal reference point, which depend strongly on environmental conditions (availability, incentive) and on what the individual has learned in the past (e.g. sexuality, exploratory drive, bonding need, emotion)
Drive hierarchy	The mechanisms that determine which drive behavior is performed at any given time; the drive hierarchy is the end result of competition among drives of different intensity (see below)
Drive intensity	In homeostatic drives, intensity depends on the degree of deviation from the reference point (see above), which in turn depends chiefly on the time since the last restoration of equilibrium, i.e. the deprivation time
Reinforcement	Increase in intensity of a drive by stimuli that raise the probability of occurrence of the preceding behavior, i.e. by instrumental learning (see p. 123); a typical reinforcing stimulus is the taste of good food during eating
Incentive	Technical term for the (learned) 'reward potential' of a stimulus; e.g. a monthly wage does not provide positive reinforcement (because it is paid too long after the actual work) but it acts as an incentive to continue performing the behavior

Definition of the term 'instinctive behavior'

Instincts require innate triggers or key stimuli (e.g. parent's beak in birds); they also require endogenous stimuli (e.g. hunger contractions of the stomach) in order for a key stimulus to be recognized. Instincts are more simply organized than drives; they are species-specific, affectively neutral and blind to consequences; under certain conditions parts of instinctive behavior are performed as vacuum behavior and displacement activities.

Thirst and hunger

This subject is treated in the following chapter on thirst and hunger

Sexuality and reproduction

This subject is treated in the chapter on reproductive physiology beginning on p. 265. On this page only a few general and pathopsychobiological aspects are considered

The basic structure of reproductive behavior comprises four phases; these must occur in sequence in one individual and each must induce the corresponding phase in another individual, in order for successful mating to result; they are:

1. **Sexual attraction:** Like all following stages, in most species this is positively influenced by the androgen level in the male animal and the estrogen level in the female; other important stimuli include the odor of the sexual organs and changes in posture and color.

2. **Appetitive behavior:** Keeps the partners together; among the appetitive reactions are 'invitations' to approach and mount, erection, vocalizations, etc.

3. **Copulatory behavior:** Triggered by appetitive behavior; in the male partner consists of intromission and orgasm with ejaculation and contraction of the pelvic musculature and penis; in the female orgasm there are uterine contractions and, likewise, contractions of the pelvic musculature; the positive feelings of orgasm are correlated with the pelvic muscle contractions.

4. **Postcopulatory behavior:** Discharge of semen in the man and uterine contractions in the woman are associated with the decline of sexual excitation; the man passes through an absolute refractory period after orgasm; in the woman multiple orgasms are possible.

Sexual dysfunctions in humans are usually mental in origin; they can also be symptoms accompanying organic disease (e.g. diabetes mellitus) or induced pharmacologically (alcohol and drug abuse, side-effect of many medicines); among the sexual dysfunctions are:

- Little or no sex drive

- Impaired sexual excitation:

  In the man: primary and secondary impotence (no or insufficient erection)

  In the woman: little excitation, no lubrication and no swelling of the outer and inner labia

- Anorgasmia: failure to reach orgasm despite normal excitation and plateau phases

- Premature ejaculation (in the man)

- Functional dyspareunia in the woman (pain during sexual intercourse with no pathophysiological basis).

Learned motivation and addiction

Drug dependence

> According to the World Health Organization, dependence is a syndrome expressed in a behavior pattern in which consumption of the drug acquires priority over other behavior patterns that previously had been valued more highly. This behavior need not be present at all times. Dependence is not absolute, but exists in different degrees. The intensity of the syndrome is measured by the forms of behavior exhibited in association with the search for and consumption of the drug, and by other, resulting forms of behavior.

Psychology of addiction: opponent process theory (two-process theory) of acquired motivation

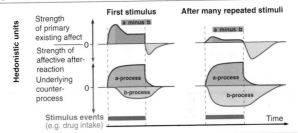

- Stimuli with a pleasurable quality first activate the pleasurably positive a-process (e.g. euphoria). Duration, intensity and quality of the a-process is directly proportional to the stimulus. After some delay, the a-process leads to opposed b-process; 'b' exhibits the reverse pleasurable quality ('withdrawal symptoms') (e.g. depression), has a longer temporal latency, increases more slowly and – this is crucial – is initially smaller than 'a' but becomes stronger with repetition (whereas 'a' always remains constant).

- Drugs with an extremely positive 'a' effect produce strong negative 'b' after-fluctuations, which can be reduced only by renewed drug intake (a), which further increments 'b' and delays the return to the neutral starting point: the vicious circle is closed. The withdrawal symptoms (abstinence phenomena, b-process) are always the opposite of the positive a-process (e.g. euphoria/depression, power/social anxiety and panic). Reduction of the negative withdrawal symptoms is the central motivator for repeated intake, not the positive reinforcing action of the drug.

- Tolerance is the situation in which the original action of the substance decreases with repeated intake (habituation); according to the two-process theory, it is a consequence of an intensification of the b-process when stimulation is repeated.

- Note also that both a- and b-processes are classically conditioned to stimuli presented shortly before or simultaneously (e.g. situation in which drugs are taken). Conditioning of the b-process, for example, causes expectation effects that also determine the degree of tolerance (e.g. the 'golden shot' is often not an overdose but the 'normal' dose in unfamiliar surroundings).

Neurobiology of addiction; example: opioid neuroadaptation (hypersensitivity theory of addiction) (after Birbaumer and Schmidt, 1996)

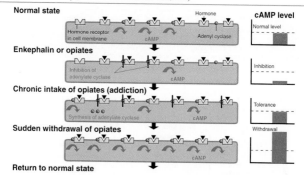

In various neurons of the pain-processing system, (primary) messenger substances (transmitters/hormones) activate the secondary messenger cAMP – which is responsible for the cell's response to receptor activation – by way of the enzyme adenylate cyclase (AC). Endogenous (e.g. enkephalins) or exogenous (e.g. heroin) opiates inhibit AC, thereby reducing the cAMP level, and ultimately produce a brief inhibition of pain. Chronic intake of opiates leads to increased AC synthesis until a normal cAMP level has been reached (development of tolerance). Upon abrupt withdrawal all AC molecules become active, so that there is an excess of cAMP which, by way of a number of intervening steps, causes the withdrawal symptoms.

Actions of various drugs; note the similarity of the effects of nicotine, cocaine, morphine and alcohol (after Birbaumer and Schmidt, 1996)

Attributes	Nicotine	Cocaine	Morphine	Alcohol	Hallucinogen	Chlorpromazine
1. Clearly distinguishable interoceptive (subjective) effects	+	+	+	+	+	+
2. Rapid rise and fall of effects	+	+	+	+	–	–
3. Pleasure closely related to dose	+	+	+	+	–	–
4. Produces euphoria	+	+	+	+	–	–
5. Acts as positive reinforcer in experiments on self-stimulation with the drug	+	+	+	+	?	–
6. Causes physiological dependence	?	?	+	+	–	–
7. Tolerance	+	+	+	+	+	+
8. Behaviorally active dose is low in comparison to damaging dose, so behaviorally active dose very frequently consumed	+	+	–	–	+	–
9. Is used socially and can determine social ranking	+	+	+	+	+	–

Basic concepts related to emotion

Definition of the terms 'emotion' and 'mood'

- Emotions are patterns of responses to stimuli inside and outside the body, which are always experienced in the dimensions (1) pleasant/unpleasant (approach/avoidance) and (2) exciting/deactivating; the distinction between emotions and motivations is only a matter of degree.

- Primary emotions are innate response patterns, which take the same form in all cultures; they include happiness or joy, sadness, fright, rage, surprise and disgust; their duration rarely exceeds seconds. The feelings of humans are usually mixtures of primary emotions.

- Moods are longer-lasting (hours, days) response tendencies, which make the occurrence of a particular emotion likely.

The 3 primary emotion systems; each system is activated by different environmental stimuli and controls corresponding behaviors (after Gray, 1982)

Emotion system	Reinforcing stimulus	Behavior
Behavioral inhibition system (BIS)	Conditioned stimuli for punishment and conditioned non-reward	Passive avoidance, extinction
Behavioral activation system (BAS)	Conditioned stimuli for punishment, reward and withdrawal of punishment	Learning to approach, active avoidance; goal-directed, cond. flight; prey aggression
Fight–flight system	Unconditioned punishment and unconditioned non-reward	Unconditioned escape response, defensive aggression

Neurobiology of avoidance (fear and anxiety, BIS)

Anatomy and physiology of the BIS

The BIS consists of the hippocampus, subiculum, entorhinal cortex (EC) and septum; three of these form a closed circuit (EC → hippocampus → subiculum → EC), the septum connects the hippocampus to the hypothalamus. The subiculum circuit has a comparison function: the arriving sensory information, preselected as significant and new, is compared with stored (expected) sensory stimulus patterns and intended movement sequences; these expectations and predictions derive from the frontothalamic system (Papez circuit); if the result of the comparison is that punishment is to be expected, the intended movement is interrupted by way of the septum (to hypothalamus) and the subiculum (to striatothalamic system).

Pharmacology of the BIS

The BIS is specifically inhibited by barbiturates, benzodiazepines (e.g. Valium) and alcohol. Therefore, of all the possible anxiety phenomena, these drugs reduce only the reactions in passive avoidance situations, e.g. frustrations, fear of innate flight stimuli ('prepared' stimuli such as snake phobia); they have no influence on unconditioned fear and aggression situations (e.g. noise) or on active avoidance (compulsive behavior, e.g. compulsive hand-washing).

Neurobiology of approach (BAS) and of aggression

Anatomy of intracranial self-stimulation (ICSS; at present the only fully developed model of positive states in the brain)

When rats are given an opportunity to stimulate themselves through electrodes implanted in the brain, by pressing a bar, many subcortical and cortical areas in the brain ('pleasure centers') are stimulated constantly by the rats (up to 5000 or more bar-presses per hour, until complete exhaustion); optimal sites are the descending medial forebrain bundle (MFB) and the lateral hypothalamus (LH); in the neocortex, the frontal cortex is especially suitable. The stimulation of lower-lying structures of the midbrain has the opposite effect: the animals try to prevent electrical stimulation of these parts of the brain ('punishment' and 'aversion' centers).

Model of dopamine activation by ICSS in the LH (modified from Stellar and Stellar, 1985)

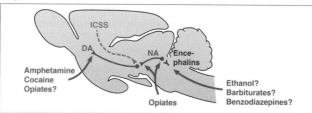

ICSS causes excitation of the dopaminergic ascending systems that function as the common final path of a descending MFB-LH system (dashed red) and of endogenous opiate neurons; pharmacological influences on the reinforcement system are shown.

Electrical stimulation of the fight–flight system in the hypothalamus, depending on stimulus site, leads to three kinds of attack behavior:

- Stimulation of the medial hypothalamus causes affective aggression; this consists in extreme attacks on the nearest accessible target, moving or not; the stimuli are aversive, and the animals soon learn bar-pressing responses to avoid them.

- Stimulation of the lateral hypothalamus causes prey aggression; the attack on prey is not aversive, has few autonomic accompaniments, consists of orderly 'cold-blooded' behavior sequences and depends on the surroundings (e.g. victim's behavior); prey can be attacked regardless of hunger, e.g. cats kill mice only for the positive reinforcement of prey aggression.

- Stimulation of the dorsal hypothalamus causes flight and flight aggression (flight attacks); the latter occurs only if during its flight the animal encounters an obstacle; on the whole, this behavior has more to do with anxiety and fright than with aggression (overlap with the BIS).

Thirst and the quenching of thirst

The following cross-references give important information about other aspects of water and electrolyte balance relevant to the survey of thirst on this and the next page:

- Water and electrolyte balance, beginning p. 260; the adequate stimuli for the regulation of water balance, the intra- and extracellular receptors and the neuronal and hormonal systems involved are the same as discussed here.

- Posterior pituitary system, beginning p. 138; there certain aspects of ADH (antidiuretic hormone, adiuretin, vasopressin) are discussed.

- Long-term regulation of the circulation, beginning p. 189; survey of the mechanisms that regulate extracellular volume and hence the filling of the vascular system.

- Thermoregulation, beginning p. 219; in particular, heat loss by the evaporation of sweat, acclimatization to heat in a tropical climate.

Origin of the feeling of thirst (from Schmidt, 1990c)

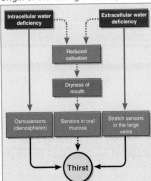

Thirst arises mainly by:

1. Decrease in cell volume due to emergence of water while the amount of salt in the cell stays constant; sensors for this osmotic thirst are osmoreceptors in the diencephalon (chiefly in and anterior to the hypothalamus).

2. Decrease in extracellular fluid; sensors for this hypovolemic thirst are stretch receptors in the walls of the large veins near the heart

- Simultaneous occurrence of 1 and 2 has an additive effect, i.e. causes especially strong thirst; on the whole, 1 is more important than 2.

- Dryness of the mouth in water deficiency is caused by decreased secretion of saliva; it is merely an accompanying symptom (e.g. moistening the mouth does not quench thirst, nor does local anesthesia there or denervation). Dryness of the mouth without water deficiency (e.g. due to speaking, smoking) produces false thirst, which can be eliminated by moistening the oral mucosa.

- Conditions that elicit thirst simultaneously cause the release of renin (and hence the formation of angiotensin II) and of ADH (antidiuretic hormone, adiuretin; see also the cross-references above).

- Thirst does not adapt; therefore as a very general rule it can be quenched only by water intake (drinking, or if necessary by gastric tube or infusion).

Mechanisms of preabsorptive and absorptive satiety (from Schmidt, 1990c)

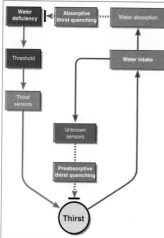

Preabsorptive satiety means that in general drinking stops before extra- and intracellular water deficiency is abolished by water absorption in the small intestine; this prevents excessive water intake, the amount drunk corresponding almost exactly to that required; the receptors and mechanisms of preabsorptive satiety are unknown.

Absorptive satiety is achieved as soon as a relative (after intake of too much NaCl) or absolute water deficit is abolished; the receptors involved ('thirst sensors') are the same ones that signal intra- or extracellular water deficiency (see preceding figure).

Thirst threshold

The water content of the human body makes up 70–75% of its weight (fat stores not taken into account). The long-term variation is only ±0.22% of the body weight, or about ±150 ml in a 70 kg man. When the body loses more than 0.5% of its weight in water, or ca. 350 ml for 70 kg body weight, the thirst threshold is reached and sensations of thirst occur. After absorptive satiety has been achieved, therefore, some time elapses before thirst recurs, despite ongoing physiological water loss; the thirst threshold thus prevents one from feeling thirst when only small amounts of water have been lost.

Primary and secondary drinking

Drinking as a consequence of thirst is primary drinking; all drinking with no obvious need for water intake is called secondary drinking. The latter is normally the usual form of liquid intake; i.e. in general we consume in advance (e.g. at mealtimes) the physiological required water, the necessary amount being estimated very precisely on the basis of previous learning and perhaps other, unknown mechanisms.

Hunger and satiety

Mechanisms of origin of the hunger sensation (from Schmidt, 1990c)

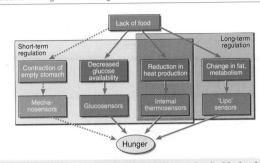

The regular daily recurrence of hunger is basically associated with the short-term regulation of food intake; in addition, there is a long-term regulation (see the two frames in the figure) that compensates dietary errors in both directions.

Mechanisms and receptors involved in eliciting hunger sensations

Concept	Description/comments
Contractions of empty stomach	The earliest presumed cause of hunger; based on the observations that (1) hunger is a general sensation localized in the stomach region and that (2) when the stomach is empty strong contractions of the stomach wall (detected by its mechanosensors) occur; however, denervation or surgical removal of the stomach has no appreciable effect on eating behavior
Glucostatic hypothesis	Decreasing availability of glucose (not the level of the blood sugar itself) is very closely correlated with the occurrence of hunger; glucose receptors are present in the diencephalon, as well as in liver, stomach and small intestine: short-term mechanism
Thermostatic hypothesis	Based on the idea that internal thermosensors (e.g. in the hypothalamus) are involved in the maintenance of an energy balance, such that a decrease in overall heat production elicits hunger; participate in short- and long-term regulation
Lipostatic hypothesis	Based on the idea that liposensors detect intermediate products of fat metabolism and evaluate these as hunger or satiety signals (in the breakdown or deposition of stored fat); suitable only for long-term regulation

Food intake without hunger. The amount of food eaten depends not only on the actual deficit and on appetite (see below), but also on the time the next meal is expected and how much energy is assumed to be spent by them (cf. secondary thirst above); in the case of sufficient food on offer, the kind of food intake will be normal

Mechanisms for preabsorptive and absorptive satiety cause the feeling of fullness (from Schmidt, 1990c)

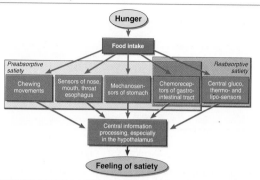

Preabsorptive satiety: Ensured by many mechnisms; in addition to the act of chewing and sensory stimuli during food intake, factors that contribute to preabsorptive satiety include stretching of the stomach and the glucose and amino acid content of the food, by way of purely sensory or with intermediate humoral steps (e.g. release of CCK).

Absorptive satiety: All three enteroceptive sensory processes associated with hunger contribute in the reverse direction to absorptive satiety (compare figure above with the preceding one); in addition involves the enteroceptive chemoreceptors that participate in preabsorptive satiety (see above).

Central mechanisms of hunger and satiety

Bilateral destruction of the ventromedial hypothalamus (VMH) induces hyperphagia (overeating), which leads to adiposity (obesity); conversely, electrical stimulation of the VMH (by means of chronically implanted electrodes) induces aphagia (refusal to eat); hence the VMH can be considered as a 'satiety center'.

Bilateral destruction of the lateral hypothalamus (LH) induces aphagia, and electrical stimulation of it causes hyperphagia; hence the LH can be considered as a 'hunger center' (for further aspects of the central nervous mechanisms of hunger and thirst see the section on motivation on pp. 127–132).

• Cross-references: the chapter on the gastrointestinal tract (GIT) considers the processes of food intake and the subsequent processes in the upper GIT (stomach, duodenum); the regulation of glucose balance is described in the chapter on endocrinology; the chapter on motivation and emotion treats the physiology of homeostatic drives (which include thirst and hunger).

General endocrinology

Classification of hormones according to chemical structure, site of action (localization of receptors) and nature of action on the target cell

Structure	Site of action	Examples/type of action/comments
Peptides	Cell membrane	E.g. insulin, ADH, many others, see p. 6; not or not very lipophilic, therefore do not permeate the plasma membrane; interaction with the membrane receptor initiates a second messenger cascade, see p. 9
Glycoproteins	Cell membrane	FSH, LH, TSH, erythropoietin; likewise not very lipophilic; interaction with the membrane receptor causes initiation of a second messenger cascade, see p. 9
Catecholamines	Cell membrane, see p. 7	(Synthesized from one molecule of tyrosine) dopamine, epinephrine and norepinephrine; are also neurotransmitters, see p. 5; depending on target cell either initiate second messenger cascades or open ion channels
Tryrosine derivatives	Cell nucleus	(Synthesized from two molecules) thyroxine (T_4) and triiodothyronine (T_3); bind to nuclear receptors to influence DNA synthesis and hence the rate of transcription of genetic information to mRNA (transcription amplification)
Steroids	Cytosol	(Cholesterol derivatives) e.g. corticosteroids; lipid-soluble, permeate the cell membrane, bind to cytoplasmic receptors and migrate with these to cell nucleus; act as transcription amplifiers, see above

Autocrine and paracrine actions of hormones: when the hormone acts on the same cells that produced it, the action is autocrine; if it acts in the immediate vicinity of the site of its release (without blood transport) it is paracrine. Typical paracrine ('tissue') hormones are the eicosanoids (prostaglandins, thromboxane, leukotrienes; from arachidonate, a polyunsaturated C_{20} fatty acid); classical hormones and neurotransmitters can also have paracrine actions.

Features and functions common to hormones in general

- Function: Can be classified according to predominant type of action: metabolic, kinetic (action on glandular secretion or pigment migration), morphogenetic or behavioral; functioning in information transfer, they are usually incorporated into control systems with negative feedback (see p. 80), which often involve central nervous structures (subject of the field of neuroendocrinology).

- Origin, secretion, transport: For most hormones these are the same as in exocrine glands (synthesis, packing in vesicles, exocytosis); exceptions are the steroid hormones, which diffuse through the cell membrane without being packed in vesicles; storage is intracellular; an exception is the extracellular storage of the thyroid hormones (see below); in the blood, hormones are often bound to carrier proteins

- Hormone titer in the blood: Kept extremely low (ca. 10^{-12} mol/l) by regulatory processes; the advantage is that when they are needed, small absolute changes amount to large relative changes; release cascades can provide considerable amplification (e.g. 0.1 µg CRH releases 1.0 µg ACTH and this in turn causes release of 50 µg of corticosteroids, i.e. amplification by a factor of 500).

Posterior pituitary (PP), neurohypophysis

Sites of synthesis and storage of ADH and oxytocin (modified after Wuttke, 1989a)

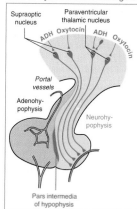

Synthesis of the high-molecular-weight precursors of the peptide hormones ADH (antidiuretic hormone or vasopressin) and oxytocin (nine amino acids each) occurs in the magnocellular nuclei (the supraoptic and paraventricular nuclei of the hypothalamus); by axonal transport in transport granules these precursors are carried to the axon terminals which form the PP, where ADH and oxytocin are enzymatically cleaved from their respective precursors.

Storage sites are the axon terminals, from which the hormones are released into the blood by exocytosis. Release is triggered by action potentials conducted along the axons. Both ADH and oxytocin are produced in both nuclei, but each individual neurosecretory cell produces and stores only one of the hormones.

Actions of ADH and oxytocin

Hormone	Action/comments
ADH	• **Antidiuretic action:** The activity of the ADH neurons is increased by plasma hyperosmolarity, and inhibited by hyposmolarity (it is possible that the ADH-producing cells are themselves osmoreceptive; in addition there are other osmosensors in the diencephalon and gastrointestinal tract; for role in thirst see p. 133); ADH increases the permeability of the renal collecting tubules to water and thereby produces antidiuresis (pp. 255, 256; mechanism of diuretic action of alcohol and pathophysiology of diabetes insipidus are also described there). • **Vasopressor action:** Injection of a relatively large amount of ADH causes arterial vasoconstriction and hence raises blood pressure; pathophysiologically large blood pressure decreases (shock, severe hypovolemia) stimulate ADH release, tending to correct blood pressure.
Oxytocin	• **Milk ejection reflex:** Mechanical stimulation of the nipples of a nursing mother (suckling) causes release of oxytocin in boluses from the PP. Oxytocin release has also been shown to occur with anticipation of suckling, e.g. in response to infant crying. Oxytocin induces contractions of the myoepithelial cells enclosing the alveoli of the mammary gland, resulting in rapid release of milk ('milk let-down'). The action of oxytocin is to induce release of secreted, stored milk. • **Ferguson's reflex:** Release of oxytocin in response to mechanical stimulation of uterus, cervix and vagina. Oxytocin stimulates myometrial contractions during labor and contributes to hemostasis after delivery. One mechanism of its contractile effect is stimulation of prostaglandin release. Estrogens upregulate oxytocin receptors and sensitize the uterus to oxytocin. Uterine oxytocin receptors reach maximum expression levels near term.

Anterior pituitary (AP), adenohypophysis

Of the six hormones secreted by the AP, 4 have as their main function to stimulate hormone secretion by other endocrine glands and 2 exert major effects on nonendocrine tissues (modified after Wuttke, 1989a)

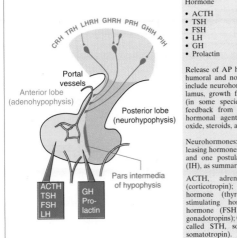

Hormone	Main site of action
• ACTH	adrenal cortex, p. 142
• TSH	thyroid gland, p. 141
• FSH	gonads, pp. 265, 266
• LH	gonads
• GH	all cells, p. 140
• Prolactin	many cells, p. 140

Release of AP hormones is controlled by humoral and non-humoral factors. These include neurohormones from the hypothalamus, growth factors, direct innervation (in some species), negative or positive feedback from target organs, and non-hormonal agents (catecholamines, nitric oxide, steroids, amino acids).

Neurohormones: There are five known releasing hormones (RH), and one identified and one postulated inhibiting hormones (IH), as summarized below.

ACTH, adrenocorticotropic hormone (corticotropin); TSH, thyroid-stimulating hormone (thyrotropin); FSH, follicle-stimulating hormone; LH, luteinizing hormone (FSH and LH are the two gonadotropins); GH, growth hormone (also called STH, somatotropic hormone or somatotropin).

Hypothalamic releasing hormones (RH) and inhibiting hormones (IH)

Abbreviation	Name	Target hormone
Releasing hormones		
CRH	Corticotropin-RH	ACTH
TRH	Thyrotropin-RH	TSH
LHRH (GnRH)	Luteinizing hormone-RH	FSH, LH
GHRH	Growth hormone-RH	GH
PRH	Prolactin-RH	Prolactin
Inhibiting hormones		
Somatostatin	Growth hormone-IH (GHIH, SS)	GH
PIH (?)	Unidentified	Prolactin

There is as yet no uniform nomenclature; the RHs are also called liberins (e.g. corticoliberin), the IHs statins (somatostatin); the original term factor is still reflected in some alternative abbreviations, such as CRF instead of CRH. Dopamine inhibits prolactin secretion and some investigators consider that it is PIH.

The hypothalamic neurons which produce RH and IH receive excitatory and inhibitory synaptic inputs from many parts of the brain, especially the mesencephalon, limbic structures and hippocampus.

Actions and regulation of GH (modified from Wuttke, 1989a)

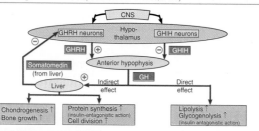

- GH is released in 3–4 pulses/day as well as in deep sleep (SS 3,4); it has diverse actions on the body's growth and metabolism, some direct and some by way of somatomedins (growth factors from the liver), e.g. somatomedin C, which stimulates protein synthesis and cell division; distinctly more GH is released in children than in adults.

- The release of GH is initiated by GHRH and inhibited by somatostatin; the somatomedins formed by GH provide negative feedback to the hypothalamus, closing the control circuit.

- The pulsatile release has a double effect on blood glucose level: first it falls due to the action of somatomedin C (previously also called IGF, insulin-like growth factor); about 1 h after GH release the direct insulin-antagonistic action takes over.

- GH deficiency in children leads to normally proportioned dwarfism; excess in children produces gigantism, and in adults acromegaly.

Actions and regulation of prolactin (modified from Wuttke, 1989a)

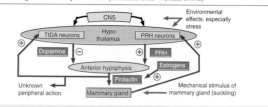

Prolactin controls the initiation and continuation of milk synthesis in the mammary gland; its RH is PRH; its IH is either an unidentified peptide or dopamine from the hypothalamic tuberoinfundibular (TIDA) neurons; prolactin has a feedback action on the TIDA neurons, closing the control circuit; other important stimuli are indicated in the figure (e.g. suckling, estrogens, environmental influences); it may be that several peptides serve as PRH (TRH, see figure opposite, VIP, angiotensin II, β-endorphin).

Thyroid gland

Hypothalamic–pituitary–thyroid control system (modified from Wuttke, 1989a)

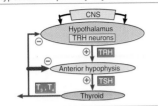

TRH from the hypothalamus releases TSH in the PP, and this stimulates the thyroid gland cells to produce and release T_3 and T_4; these exert negative feedback control on the hypothalamus and PP; this closes the control circuit.

Production (synthesis), storage and release (secretion) of T_3 and T_4 occur in a sequence of seven steps

1. The cells of the thyroid gland produce thyroglobulin, a high-molecular-weight protein containing many tyrosine molecules.

2. Within the thyroglobulin each tyrosine molecule binds 1 or 2 iodine atoms (iodine requirement in food 150 µg/d).

3. The combination of 2 iodotyrosine molecules forms either triiodothyronine (T_3) or tetraiodothyronine (T_4, thyroxine, formed in larger amounts).

4. The thyroglobulin, including the T_3 and T_4 incorporated in it, is temporarily stored extracellularly in the thyroid gland follicles (several months' supply).

5. In order to release T_3 and T_4 thyroglobulin is taken back into the gland cells by pinocytosis and broken down.

6. T_3 and T_4 diffuse into the blood under the influence of TSH (see above) and become bound to plasma proteins, mainly in non-covalent form.

7. In the blood ca. 30% of the T_4 is deiodinated to form T_3 (i.e. 80–90% of the T_3 is formed outside the thyroid gland).

Hormonal effectiveness of T_3, T_4 and rT_3

T_3 (Triiodothyronine) is the form of thyroid hormone that is metabolically active on most cells of the body.

T_4 (Tetraiodothyronine, thyroxine) is largely ineffective metabolically, but it contributes to the negative feedback in the control system (see figure above).

rT_3 (Reverse T_3) is formed mainly outside the thyroid gland by deiodination at the 'wrong' position, i.e. at the phenol rather than the tyrosine ring; it is biologically inactive.

Cellular and systemic actions of T_3 (and T_4)

As stated on p. 132, T_3 and T_4 have their receptors in the nucleus and enhance transcription, thereby stimulating the synthesis of proteins (e.g. enzymes) in all cells of the body; in addition they act to increase enzyme activities and stimulate the Na,K pump; all of this causes an increase in energy metabolism (metabolic variables and measurement pp. 217, 218).
• Hypothyroidism: inadequate thyroid function; reduced basal metabolism
• Hyperthyroidism: excessive thyroid function; increased basal metabolism

Adrenal cortex

Layers of the adrenal cortex (AC) and their hormones

Layer	Hormones	Comments
Zona glomerulosa (outer layer)	Mineralocorticoids	Main representative is aldosterone; its main action is to enhance Na⁺ reabsorption in the distal tubule and collecting tubule of the kidney, see next page.
Zona fasciculata (middle layer)	Glucocorticoids	Main representative is cortisol; its main action is gluconeogenesis (protein catabolic action), see below
Zona reticularis (inner layer)	Androgens	Chief androgen source in women, in men provides $^1/_3$ of the androgens, $^2/_3$ coming from the testes; main representative here is DHEA dehydroepiandrosterone, see p. 265

Hypothalamic–pituitary–adrenal control system (modified from Wuttke, 1989a)

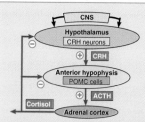

CRH from the hypothalamus elicits release of ACTH in the AP and this stimulates the cells of the AC to produce and release cortisol, which exerts negative feedback control on hypothalamus and AP, closing the control circuit; the control system exhibits a clear circadian rhythm with maximal cortisol level in the morning, minimal at night; the release is episodic (pulsatile).

• Stress situations can alter the function of the control system due to increased release of CRH; as a result, the ACTH level can become higher than necessary for maximal cortisol secretion.

• ACTH is produced in the AP in cells that, from the same precursor, also produce β-endorphin and α-melanocyte-stimulating hormone (α-MSH). All of these peptides are contained in the same precursor, namely proopiomelanocortin.

Actions of cortisol (and other glucocorticoids)

• **Metabolic actions:** Stimulation of gluconeogenesis in the liver from amino acids detained from breakdown of muscle proteins (protein catabolic action); at the same time synthesis of muscle proteins is reduced (antianabolic action); glucose transport into the cells and utilization of glucose are reduced; the fatty acid levels in the blood are raised due to degradation of triglycerides.

• **Permissive action:** Sensitization of vascular smooth muscle to catecholamines; the result is that during stress the blood is redistributed to skeletal muscle.

• **Anti-inflammatory (immunological) actions:** Inhibition of inflammation and immunosuppression only at relatively high (pharmacological) concentrations.

• **Psychological actions:** Psychological and behavioral changes (euphoria or depression) are produced by excess and deficit, respectively.

Mineralocorticoids; control and actions of aldosterone

Aspect	Description/comments
Biochemistry	Biosynthesis from cholesterol (provided by blood plasma or formed in the AC; applies to all corticosteroids); ACTH simulates biosynthesis of aldosterone; aldosterone is excreted in bile and urine after conjugation to glucuronic acid in the liver
Control of release	(1) Reduced extracellular volume, (2) hyperkalemia and (3) ACTH. (2) and (3) act directly on the cells of the zona glomerulosa; (1) (the main factor controlling aldosterone secretion) causes an increase in circulating levels of angiotensin II (renin–angiotensin–aldosterone mechanism). Angiotensin II stimulates aldosterone secretion by the adrenal gland
Actions	In collecting tubule, aldosterone stimulates (1) Na^+ reabsorption and (2) K^+ secretion (see pp. 255, 258); secondarily, if ADH is present, water reabsorption increases. These effects tend to increase extracellular fluid volume, counteract the stimuli that initiated release, and close the control circuit

The onset of action of aldosterone is slow (0.5–1 h after release or injection) and the maximal effect is not reached for several hours, because of the many intermediate steps between binding to its cytosolic receptor (p. 137) and the synthesis of the proteins that mediate its cellular actions (the same applies to all corticosteroids).

Hormones of the adrenal medulla (AM)

Synthesis of catecholamines, storage in AM, peptidergic co-transmitters, innervation of the AM, release, sites and nature of actions

- **Biochemistry**: The cells of the AM produce (from tyrosine) and store the catecholamines epinephrine (E, 80%) and norepinephrine (NE, 20%); E and NE are stored in various cells of the AM; co-transmitters in the AM cells are substance P, VIP, somatostatin, CCK, metenkephalin (see also p. 6)

- **Innervation**: AM cells are embryologically homologous to postganglionic neurons; therefore they have a preganglionic cholinergic innervation, see Fig. p. 103

- **Release** of E and NE is exclusively neuronal; for conditions promoting release, see p. 104

- **Sites of action**: In principle the same target organs as those of postganglionic sympathetic neurons (see p. 102), especially effectors with little or no postganglionic innervation (e.g. media of arteries, hepatocytes and adipocytes)

- **Metabolic actions**: Main actions of E and NE from AM is on hepatocytes, adipocytes and myocytes (see p. 102); glycogen breakdown and gluconeogenesis result in increased glucose supply; E and NE also stimulate lipolysis

Pancreatic hormones

Cell types in islets of Langerhans and the hormones they produce

Cell type	Hormone	Comments
A-cell	Glucagon	About 25% of the islet cells; glucagon is a peptide comprising 28 amino acids; cf. p. 137; peptides bind to membrane receptors and exert their effects via second messengers; in this case, cAMP
B-cell	Insulin	About 60% of the islet cells; insulin is a peptide consisting of an A-chain with 21 amino acids and a B-chain with 30 amino acids, linked together by disulfide bridges
D-cell	Somatostatin	About 15% of the islet cells; somatostatin (GHIH) is a 14-amino acid peptide; first discovered in hypothalamus; for actions see pp. 139, 140

Factors stimulating insulin secretion (modified from Wuttke, 1989a)

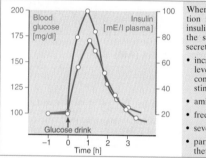

When the blood glucose concentration is normal, 70–100 mg/dl, the insulin level in plasma is very low; the strongest stimulus for insulin secretion is an

- increase in the blood glucose level (after drinking a solution containing 100 g glucose); other stimuli are
- amino acids,
- free fatty acids,
- several gastrointestinal peptides,
- parasympathetic activity (sympathetic inhibits, see p. 102)

Main actions of insulin (in roughly the order of their physiological significance)

- Increased uptake and consumption of glucose in practically all cells of the body.

- Conversion of glucose to glycogen in hepatocytes combined with inhibition of glycogenolysis.

- Promotion of amino acid uptake into cells and increase in protein synthesis (action comparable to that of GH, see p. 140).

- Conversion of glucose to fat in hepatocytes, and also in adipocytes; this occurs as soon as the available glycogen stores have been filled up, with simultaneous inhibition of lipolysis.

- Increased membrane permeability of myocytes to glucose (elevated number of carrier molecules); the myocytes then form small amounts of glycogen; during intense exercise, the plasma membrane of myocytes can also undergo an increase in glucose permeability independently of insulin (mechanism unknown).

Actions of glucagon (from Wuttke, 1989a)

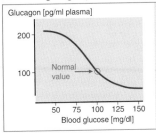

Glucagon [pg/ml plasma]

200 —

100 —

Normal value →

50 75 100 125 150

Blood glucose [mg/dl]

Under normal conditions and in hyperglycemia the blood glucagon concentration is low, in hypoglycemia it rises; other stimuli are catecholamines and amino acids.

The basic action of glucagon is antagonistic to insulin and consists of promoting glycogenolysis in the liver, hence raising the glycemia.

Actions of somatostatin from the D cells (see also p. 140)

Somatostatin has an exclusively paracrine action in the islets of Langerhans, i.e. to inhibit the secretion of insulin and glucagon; its endocrine action is to inhibit both motility and secretion in the GI tract; this slows digestion and absorption of nutrients, which prevents excessive fluctuation of the glycemia.

Stimuli for pancreatic release of somatostatin are high plasma concentrations of amino acids, fatty acids and glucose.

(For indirect actions of the antagonist GHRH on blood glucose level see p. 140; actions of glucocorticoids, primarily cortisol, on p. 142.)

Main symptoms of diabetes mellitus are polydipsia, polyuria and polyphagia; they are accompanied by hypoglycemia and glycosuria. The cause is insufficient insulin action; types of diabetes include (modified from Söling, 1991)

I. Primary (idiopathic) forms of diabetes

  A. Type I:* Insulin-dependent 'juvenile diabetes'

  B. Type II: Non-insulin-dependent diabetes. Most of them are adult-onset, i.e. maturity-onset diabetes mellitus (MOD). Maturity-onset diabetes of the young (MODY) is less frequent

II. Secondary forms of diabetes

  A. Pancreatic diabetes

  B. 'Endocrine' forms (insulin-antagonistic hormones)

  C. Iatrogenic (medication-induced) form

  D. Diabetes associated with chronic diseases

III. Diabetes due to insulin receptor disorder

  A. Insulin-resistant diabetes caused by receptor autoantibodies*

  B. Lipodystrophic diabetes

IV. Diabetes of insulin autoimmune syndrome* (Hirata syndrome)

* Autoimmune mechanisms probably involved in pathogenesis.

Cross-references for the description of other hormone systems in this book:

- Hormones and possible hormone of the GI tract, see p. 241.
- Ca^{2+} and phosphate balance, see p. 264.
- Sex hormones, see p. 265.

Composition of the blood

Normal values for the blood of adults. Cell counts and plasma values in anticoagulated blood; serum values in clotted blood

Term	Value	Comments
Total volume	4–6 liters	About 6–8% of body weight; measured by dilution of indicators, see p. 260; for distribution in the cardiovascular system, see p. 175
Hematocrit	♂ 0.42–0.52 ♀ 0.37–0.47	Erythrocyte volume expressed as fraction of blood volume; also given as vol. % (ml cells/dl blood)
Erythrocytes	♂ 4.5–6.0 ♀ 4.0–5.2	(Million/µl blood); measured by counting a standard volume of blood diluted 100- to 200-fold; for properties and functions of erythrocytes see p. 150
Reticulocytes	5–10‰	(‰ of erythrocytes) Index of erythropoietic activity, see p. 150
Leukocytes	4000–10 000	(Number per µl blood) Measured as above; not a uniform group; for properties and functions see p. 152.
Thrombocytes (platelets)	150 000–400 000	(Number per µl blood); important for hemostasis, see p. 155
Plasma	(2.25–3.35 liters)	Blood volume minus red cell volume (all other cellular constituents have negligibly small total volumes); for total body water see p. 260
Serum	Like plasma	Plasma without fibrinogen (extracted during clotting, q.v.)
Total protein	6.0–8.0 g/dl	Plasma (protein concentration)
Albumin	3.5–5.0 g/dl	(ca. 60% of total protein); accounts for most of the plasma colloid osmotic pressure (MW 69 000), also transport protein for many substances, e.g. T_3, T_4
Globulins	2.2–4.0 g/dl	(ca. 40% of total protein), electrophoresis shows 4 subfractions, see below; each can be further subdivided by immunoelectrophoresis
Viscosity (rel. to $H_2O = 1$)	Blood 4.5 Plasma 2.2	Viscosity of blood increases with increasing hematocrit; for importance to flow resistance see p. 177; plasma viscosity is determined mainly by protein content
Osmolality	290	[mosmol/kgH_2O]; measure of the total number of particles dissolved in plasma (96% are electrolytes); measured by freezing point depression or vapor pressure osmometer
		Plasma osmolality accounts for a plasma osmotic pressure of 745 kPa (7.3 atm, 5600 mmHg). This is the (osmotic) pressure difference between plasma and pure H_2O, if they were separated by an ideal semipermeable membrane
Tonicity	1 (relative to plasma)	'Effective' osmotic pressure (Osm_{eff}) of a solution relative to plasma: isotonic = 1, hypertonic >1, hypotonic <1, e.g. a solution of 0.9 g NaCl in 1 liter H_2O is isotonic ($Osm_{eff} = \sigma_s \cdot Osm$, where Osm is the total osmolality and σ_s is the solute reflection coefficient, denoting solute permeability relative to water; $\sigma_s = 1$ for impermeant solute, $\sigma_s = 0$ for solute as permeable as water)
Oncotic pressure	3.3 kPa (25 mmHg)	(Syn.: colloid osmotic pressure); exerted between plasma (high protein concentration) and interstitial fluid (low protein concentration). Most capillary walls have a very low permeability to proteins

Components of the plasma and its functions

Average concentrations of electrolytes and non-electrolytes in human plasma
(modified from Weiss and Jelkmann, 1989)

Kind of particle	meq/l	mmol/kg plasma water
Electrolytes		
Cations		
Sodium	143	153
Potassium	5	5
Calcium	5	3
Magnesium	2	1
Total	155	
Anions		
Chloride	103	110
Bicarbonate	27	28
Phosphate	2	1
Sulfate	1	1
Organic acids	6	
Protein	16	≈ 1
Total	155	
Non-electrolytes	g/dl	
Glucose	0.9–1.0	5
Urea	0.40	7

Protein electrophoresis of human serum (modified from Weiss and Jelkmann, 1989)

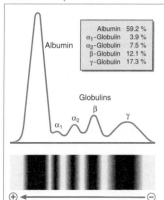

Albumin	59.2 %
α_1-Globulin	3.9 %
α_2-Globulin	7.5 %
β-Globulin	12.1 %
γ-Globulin	17.3 %

Electrophoresis is the migration of electrically charged solute particles in a steady electric field; Serum protein electrophoresis is an important diagnostic aid, because many diseases (e.g. infections, malignancies, nephrotic syndrome) produce characteristic changes; combined with immunoprecipitation (immunoelectrophoresis) it allows separation of additional components; for normal values and functional significance of the individual fractions see opposite page and p. 157.

Normally in 24 h ca. 17 g albumin and 5 g globulin are synthesized de novo in the liver; the half-life of albumin is 10–15 d, that of globulin ca. 5 d.

Protein fractions in human blood plasma; MW, molecular weight; IP, isoelectric point (data from many authors, assembled by Weiss and Jelkmann, 1989)

Protein fraction / Electrophoretic	Immunoelectrophoretic	Mean concentration g/dl	Mean concentration µmol/l	MW (kDa)	IP	Physiological significance
Albumin	Prealbumin	0.03	4.9	61	4.7	Binding and transport of thyroxine and other substances; colloid osmotic pressure, body amino acid pool
	Albumin	4.00	579.0	69	44.9	
α_1-globulins	Acidic α_1-glycoprotein	0.08	18.2	44	2.7	Product of tissue degeneration
	α_1-lipoprotein ('high-density lipoproteins')	0.35	17.5	200	5.1	Lipid transport
α_2-globulins	Ceruloplasmin	0.03	1.9	160	4.4	Oxidase activity, copper transport
	α_2-macroglobulin	0.25	3.1	820	5.4	Plasmin and proteinase inhibition
	α_2-haptoglobin	0.10	11.8	85	4.1	Binds free hemoglobin; prevents loss in urine
β-globulins	Transferrin	0.30	33.3	90	5.8	Iron transport
	β-lipoprotein ('low-density lipoproteins')	0.55	0.3–1.8	3000–20000	–	Lipid transport (esp. cholesterol)
	Fibrinogen	0.30	8.8	340	5.8	Blood clotting
γ-globulins	IgG	1.20	76.9	156	5.8	Immunoglobulins: see immunology texts
	IgA	0.24	16.0	150	7.3	
	IgM	0.12	1.3	960		
	IgE	0.00003	0.002	190	–	

Properties and functions of erythrocytes

Main function of erythrocytes

Transport of O_2 and CO_2 between lungs and tissues; see chapter on respiratory gas transport and acid–base status of the blood, beginning p. 204

Shape and size of the erythrocytes (for number of erythrocytes see p. 147)

Non-nucleated red blood cells are biconcave disks of ca. 7 μm diameter; the flattened shape makes the surface large relative to the volume of 80–100 μm³; the erythrocytes are extremely deformable; this plasticity is the basis of the relatively low viscosity of blood flowing in small vessels; during slow flow, however, they aggregate (rouleaux formation, see p. 177).

The half-life of normal erythrocytes is 110–120 d; aged erythrocytes are removed from the blood and destroyed, mainly in the spleen

Erythrocyte sedimentation rate (ESR) (measured by method of Westergren)

Because erythrocytes are heavier than plasma (specific gravity 1.096 and 1.027, respectively), they slowly sink ('sediment') in blood that has been prevented from clotting and is allowed to stand. The normal ESR of a blood column 10 cm long, after 1 h, is 3–6 mm for men and 8–10 mm for women; ESR increases with greater aggregation; this in turn is promoted by disease-induced globulin changes (e.g. in inflammation); hence an elevated ESR suggests abnormalities in plasma proteins.

Erythropoiesis (formation of new erythrocytes)

• In adults, it occurs exclusively in the red bone marrow; formed from pluripotent stem cells, initially by way of ca. 4–7 intermediate steps in the marrow; the last nucleated progenitor is the normoblast; after enucleation, which converts it to a reticulocyte, the cell emerges through the slits in the marrow sinus and enters the blood.

• 0.5–1% of the erythrocytes in the blood are reticulocytes (see p. 147); there, within 1–2 d, they lose mitochondria and ribosomes, alter their membrane glycoproteins, and finally mature; after blood loss or when the erythrocyte life-span is pathologically shortened, the rate of erythropoiesis is elevated; the number of reticulocytes in the blood rises correspondingly.

• Erythropoiesis is controlled by the hormone erythropoietin which is released by the renal glomeruli (also 10% in the liver) in response to falls of O_2 partial pressure in arterial blood or reduced blood flow to the kidneys, e.g. in acclimation to high altitudes (see p. 234) and after loss of blood.

Human blood groups

ABO system

Blood group (phenotype)	Genotype	Antigens on erythrocytes	Antibodies in plasma	Frequency in USA
O	OO	None	Anti-A Anti-B	46%
A	AA AO	A	Anti-B	42%
B	BB BO	B	Anti-A	9%
AB	AB	A and B	None	3%

Explanation of the ABO system

- Assignment to a blood group depends on the antigens A and B (agglutinogens, chemically glycoproteins) in the erythrocyte membrane, see Table; the antibodies in the plasma (agglutinins) are formed during the 1st year of life, against those antigens not carried by the child's own erythrocytes.

- The blood group properties A and B are dominant over O, so that the phenotype O occurs only in homozygous form (genotype OO). Parents with phenotype blood groups A and B (with genotypes AO and BO) can therefore have children with blood of group O; A and B are co-dominant (phenotype AB).

- Anti-A and anti-B are IgM antibodies (isoagglutinins of the β-globulin fraction); when they contact their antigens the complement system is activated, leading to agglutination of erythrocytes and/or hemolysis with potentially lethal consequences.

Determination of blood group in the ABO system by means of test serums
(after Wintrobe, 1981)

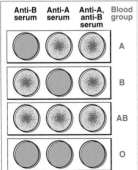

- The erythrocytes to be tested are mixed on a glass slide with three commercially available antiserums; the resulting patterns of agglutination (clumping of erythrocytes) unequivocally identify the blood group.

- Conversely, serum can be tested by mixing it with erythrocytes of known blood group.

- Before transfusions it is customary to perform additional cross-matching of donor erythrocytes with recipient serum (major test) and of recipient erythrocytes with donor serum (minor test), to rule out errors (wrong samples, poor observation) and incompatibilities other than those of the ABO system.

Rh (Rhesus) system

- There are six Rh antigens in the erythrocyte membrane: C, D, E, c, d, e. Of these, only D has such a strong antigenic action that it plays a clinical role; 85% of humans have D and are called Rh-positive or Rh; the remaining 15% are Rh-negative (rh); in the phenotype Rh the genotypes are DD or Dd, in the phenotype rh it is always dd.
- Anti-D antibodies are formed (within a few months) by rh individuals only after exposure to D; this occurs only (1) after transfusion of Rh blood to rh people or (2) when D erythrocytes from a child pass through to the placenta to an rh mother (most likely to happen during birth).
- Hence the first wrong transfusion and the first pregnancy of an rh mother with an Rh fetus cause no problems (sensitization only), but subsequent ones induce agglutination in the patient or in the fetus.
- Anti-D prophylaxis: by injection of an anti-D γ-globulin into the rh mother immediately after birth/miscarriage, the Rh erythrocytes acquired from the child are destroyed; then the mother forms no anti-D antibodies.

Additional erythrocyte antigens

At least a dozen blood group systems in addition to those above (e.g. Kell, Duffy, MNSs) have been described. They are occasionally of clinical relevance, especially in difficult cross-matching, and also used in forensic medicine.

Properties and functions of leukocytes

Differential normal distribution (leukocytes per μl blood and in % of total)

Type of cell	Adults	
	Absolute	%
Immature	0	0
Granulocytes		
Neutrophils segmented	1500–7500	34–75
band nuclei	0–800	0–8
Eosinophils	0–500	0–5
Basophils	0–300	0–3
Lymphocytes	500–5000	12–50
Monocytes	100–1500	3–15
Total leukocytes	4000–10000	100

- Absolute leukocyte counts in children and adolescents up to age 14 are distinctly higher than those of adults, e.g. newborns 9000–30000/μl, 1 month–3 years 5500–18000/μl, 4–7 years 5500–15500/μl, 8–13 years 4500–13500/μl.
- Absolute leukocyte counts in African-Americans are ca. 15% lower than in Caucasians.
- Leukocytosis: >10000 leukocytes/μl blood in adults; observed in many diseases, e.g. infections; extreme in leukemias.
- Leukopenia: <4000 leukocytes/μl blood in adults; involves mainly the neutrophils, extreme form is agranulocytosis.
- Leftward shift: Appearance of immature forms and greater numbers of band neutrophils in many diseases, see figure on next page.

Distribution of neutrophils in health and in various diseases
(modified after Haden, 1935)

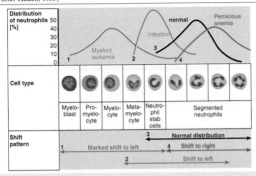

From left to right progressive maturation and finally aging of neutrophils are depicted; the figure illustrates the terms leftward and rightward shift and shows three typical syndromes with which they are associated.

Features and functions common to all leukocytes

Concept	Description/comments
Myelopoiesis	All leukocytes, together with erythrocytes and thrombocytes, descend from the same pluripotent hemopoietic stem cells; differentiation is controlled by growth factors (e.g. CSF, colony-stimulating factors for granulocytes, interleukin-2 for lymphocytes)
Life-span	Non-uniform, from a few hours to 100–300 d; however, time spent in blood is often only 1–2 h, rarely more than 24 h, i.e. there is continual exchange from the blood into various tissues and back (see migration)
Margination	A fraction of the granulocytes (marginated pool) is reversibly attached to the vascular endothelium; the remainder circulates rapidly (circulating pool); the marginated pool can be mobilized, e.g. by cortisol and catecholamines (work, stress), producing pseudoleukocytosis.
Migration	For the granulo- and lymphocytes the blood is a transport organ, carrying them from their sites of origin to target sites; >50% of the leukocytes are located in the interstitial space, >30% in the bone marrow; the process of crossing the capillary wall is called leukodiapedesis.
Phagocytosis	Foreign particles (e.g. bacterial) 'attract' leukocytes, especially neutrophils, by chemotaxis (elicited by chemokines). The leukocytes migrate and then endocytose the particles. Pus is a mixture of dead leukocytes, tissue debris, bacteria, etc.

Properties and functions of individual types of leukocytes

Leukocytes	Property/function/comments
Neutrophils	Cell diameter (also of other granulocytes) 10–17 µm; most important agent of 'non-specific defense system' chemotaxis, phagocytosis. Granules are rich in proteases and myeloperoxidase. Activated neutrophils release arachidonic acid for synthesis of eicosanoids (see p. 186)
Eosinophils	Granules contain eosinophil peroxidase and cationic protein, cytokines, etc. Cell count varies in inverse proportion to circadian periodicity of glucocorticoid level, hence is highest at midnight (see p. 142); eosinophilia is most common in allergic, especially autoimmune diseases, drug reactions and parasitic infections
Basophils	Granules contain heparin and histamine; involved in allergic reactions; also, by releasing the platelet-activating factor (PAF), are involved in the activation and aggregation of thrombocytes
Monocytes	Independent leukocyte form with agranular cytoplasm; diameter 12–20 µm; best phagocytosis capacity; migrate and after final maturation form the tissue macrophages (histiocytes); producers of cytokines, interleukin-1, etc.
Lymphocytes	Their progenitor cells migrate from the bone marrow to mature in the lymphatic organs; increase to >5000/µl in adults is called lymphocytosis, abnormally low counts are lymphopenia; agents of the specific defense system, see below

Hemostasis, clotting, fibrinolysis

Summary of the processes involved in hemostasis and clotting (coagulation) (after Schmidt, 1983)

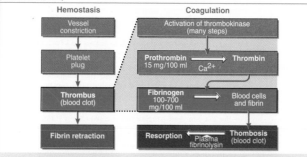

The figure is highly schematic; it shows that blood clotting is a subprocess in hemostasis, the mechanism that stops bleeding (see dotted lines); the thrombus (blood clot) consists of a network of coagulated fibrin together with the enclosed blood cells; this basic diagram of clotting goes back to Morawitz (1905); for a more detailed survey see next page.

Scheme of clotting and fibrinolysis (modified after Weiss and Jelkmann, 1989, assembled by them from the publications of various groups)

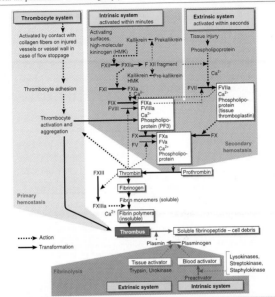

- **Thrombocytes:** After activation from the resting state (mainly by surface contact and some clotting factors, e.g. thrombin) they play crucial roles in several stages of hemostasis, in particular the initial ones, by releasing their contents and by components in their membranes (e.g. platelet factor 3, PF3)

- **Primary hemostasis** (left): Preliminary stage (see p. 154), in which bleeding is stopped chiefly by vasoconstriction and a thrombocyte plug.

- **Secondary hemostasis** (middle and right): Cascade of the processes leading to thrombus formation; has the following two starting points:

- **Intrinsic system** (middle): Process of clotting begins with clotting factors in the plasma itself.

- **Extrinsic system** (right): Process of clotting begins with phospholipoproteins released from damaged cells.

- **Clotting factors** (see p. 156): Substances involved in clotting, mostly proenzymes; the activated proteolytic enzymes are identified by the suffix a (e.g. II prothrombin, IIa thrombin)

Clotting factors

Factor	Synonym	Comments
I	Fibrinogen	Soluble protein, fibrin precursor
II	Prothrombin	α_1-Globulin, proenzyme of thrombin
III	Tissue thromboplastin	Phospholipoprotein, acts in extrinsic system
IV	Calcium ions	Required for activation of several other factors
V	Proaccelerin	β-globulin, Va is part of prothrombin activator
VII	Proconvertin	α-globulin, proenzyme, VIIa is protease
VIII	Antihemophilic factor	Deficit: classical hemophilia
IX	Christmas factor	Deficit: hemophilia B
X	Stuart factor	α_1-Globulin, proenzyme, Xa is protease
XI	PTA	(Plasma thromboplastin precursor or antecedent)
XII	Hagemann factor	β-globulin, proenzyme, XIIa is protease
XIII	Fibrin-stabilizing factor	β-globulin, XIIIa causes cross-linking of fibrin

The von Willebrand factor (vWF) appears to circulate complexed with factor VIII. It promotes adhesion of platelets to each other and to the vascular endothelium. Its deficiency (von Willebrand's disease) is the most common bleeding disorder diagnosed in adults.

The physiological intravascular inhibition of clotting (prevention of spontaneous thromboses in vascular system) depends primarily on:

1. **Antithrombogenic property of normal vascular endothelium:** Surface molecules (endothelial glycocalyx) prevent both attachment of thrombocytes and activation of contact-sensitive clotting factors; these are primarily • antithrombin III (inhibits IIa, IXa–XIIa), • α_2-macroglobulin (inhibits IIa, kallikrein, plasmin) and • protein C (inhibits Va, VIIIa); the endothelium itself also secretes antithrombogenic substances, mainly • adenosine, • prostacyclin (PGI$_2$), • EDRF (see p. 186).
2. **Blood flow:** When blood flows slowly, effective concentrations of activated clotting factors can be more easily produced at defects in the vessel walls; hence rapid blood flow is antithrombogenic.
3. **Fibrinolysis:** Operates continually in the normal vascular system, counteracting the continual slight conversion of fibrinogen into fibrin (for steps in plasmin formation see lower part of figure on p. 155).

For the therapeutic inhibition of clotting the following anticoagulants are used:

- **Heparin;** is also produced by the body itself; enhances the action of antithrombin III, inactivates IIa, IX, X, thus generally inhibits the formation and action of thrombin; must be administered parenterally; its effect is of short duration (4–6 h); antidote: protamine.
- **Coumarins** act as vitamin K antagonists in the liver, inhibiting the synthesis of factors II (prothrombin), VII, IX, X; can be administered orally; long-lasting action; antidote: vitamin K.
- **Aspirin** has an antithrombotic effect by inhibiting platelet function (inhibition of cyclooxygenase).

For therapeutic fibrinolysis the following fibrinolytics are used:

- **Urokinase;** also found in urine, see figure on p. 155; promotes the conversion of plasminogen to plasmin; antifibrinolytics have the opposite action.
- **Streptokinase;** derived from hemolytic streptococci; action like that of urokinase; like urokinase, useful for the treatment of thrombosis.

Some tests for evaluation of hemostasis

Test	Norm	Test function/comments
Thrombocyte count	150 000–400 000/µl	Pathological finding in thrombocytosis or thrombocytopenia
Bleeding time	laboratory-dependent	Test of platelet function
Prothrombin time (PT)	10–12 s	Screening test of the extrinsic system, monitoring coumarin therapy
Partial thromboplastin time (PTT)	40–50 s	Screening test of the intrinsic system, monitoring coumarin therapy
Thrombin time (TT)	17–25 s	Monitoring heparin therapy

In addition, specific assays measure concentration or activity of specific factors.

The International Normalized Ratio (INR = measured PT/standard PT) is used to standardize PT determinations during anticoagulant therapy.

For defense functions of the blood, see immunology texts.

Basic mechanisms of cardiac excitation

Myocardial cells are electrically coupled, forming a functional syncytium

The heart muscle consists of interwoven myocardial cells, joined at their boundaries by intercalated disks, which contain gap junctions. These junctions permit cell-to-cell conduction of excitation; hence the myocardial cells form a functional syncytium. As a rule, excitation spreads over the entire myocardium (except that the AV node usually does not conduct a ventricular extrasystole back to the atria, see below).

Most myocardial cells are part of the contractile myocardium, but a small number constitute the specific excitation–conduction system

1. The mechanical pumping work of the heart is performed by the contractile myocardium of the atria and ventricles.

2. Excitation is generated and rapidly propagated by cells of the excitation–contraction (EC) system (see figure on p. 160). They include (a) sinus node (SA node), (b) atrioventricular node (AV node), (c) bundle of His, (d) right and left bundle branches, and (e) Purkinje fibers.

Muscle fibers of the EC system have fewer fibrils and more sarcoplasm than contractile fibers, and contain fewer mitochondria. The bundle-branch and Purkinje fibers have very large diameters (and hence high conduction velocity, see below).

Ionic mechanisms of resting potential (RP) and action potential (AP) of contractile cells (atrial and ventricular myocardium)

Term	Definition/comments
Resting potential	Ranges from about -70 to about $-80\,mV$. It is close to the K^+ equilibrium potential; a small part, at most $10\,mV$, is contributed by the electrogenic Na^+, K^+-ATPase (net transport of one positive charge out of the cell per cycle). Unless the contractile cells are excited, their membrane potentials remain at the resting level.
Cardiac action potential: description and ionic mechanisms (action potential duration $200–400\,ms$)	The cardiac action potential is divided into five phases: Phase 0: the membrane voltage changes from ca. -80 to ca. $+30\,mV$ in $1–2\,ms$ (inversion of the sign of the voltage is called overshoot). The mechanism is a rapid, large increase in Na^+ conductance (g_{Na}), causing a large inward current; the channel has a threshold for activation of $-60\,mV$ and is TTX-sensitive. Phase 1: the membrane voltage repolarizes a few mV. The main mechanism is the beginning of the rapid inactivation of Na^+ channels (concomitantly there are increases in g_{Ca} and g_K, so that the net change in inward current is small). Phase 2: Corresponds to the plateau, i.e. sustained depolarization at a level slightly lower than the peak depolarization. The main mechanism is the activation of g_{Ca} (sustained inward current, threshold ca. $-30\,mV$, blocked by Ca^{2+} antagonists such as verapamil and nifedipine), followed by its slow inactivation. Phase 3: Rapid repolarization. This phase is the result of complete inactivation of g_{Na} (rapid) and g_{Ca} (slower) and rapid activation of g_K (outward current). Phase 4: 'Resting' membrane voltage, dominated by g_K (constant in most myocardial cells, undergoes spontaneous depolarization in pacemaker cells).

During the AP plateau the heart is inexcitable (absolute refractory period). Excitability returns gradually to normal during repolarization (relative refractory period)

> The refractory period is attributable to the inactivation of the TTX-sensitive Na^+ channels by the maintained depolarization during the plateau phase. Recovery begins as soon as the membrane potential has returned to ca. $-40\,mV$. Hence, the refractory period is correlated with the duration of the AP. The refractory period protects the heart from being re-excited too soon, e.g. by re-entry, so that it does not contract when empty.

Pacemaker and EC cells are automatic, i.e. can generate excitation by spontaneous slow, non-propagated diastolic depolarization. Actual pacemakers are the cells that depolarize most rapidly; the others are potential pacemakers

Term	Definition/comments
Sinus node (SA node)	Located in the right atrium near the opening of the superior vena cava. Is the normal, primary pacemaker. Resting frequency in adults ca. 60–80/min. Excitation spreads to both atria. AP of the sinus node cells has a slow phase 0, small amplitude and short duration. Phase 0 is largely determined by activation of g_{Ca}.
Atrioventricular node (AV node)	Only normal path for transmission of excitation from atria to ventricles; slow conduction because the fibers are thin; allows time for atrial contraction to propel blood into the ventricles before they contract. The AV node is a secondary pacemaker (typical frequency, 40–60/min).
Bundle of His, bundle branches, Purkinje fibers	Excitation–conduction system in the ventricular septum and thence into the ventricular myocardium. Conduction is rapid (ca. 2 m/s), so excitation reaches the whole myocardium nearly simultaneously; final conduction there at ca. 1 m/s. The conducting system is a tertiary pacemaker, frequency 30–40/min (ventricular–escape rhythm)
Ionic mechanisms	The maximal diastolic potential of the sinus node immediately after repolarization from the preceding AP is ca. $-60\,mV$. The pacemaker potential (diastolic depolarization) begins immediately and at $-40\,mV$ generates the next AP. Main mechanism of spontaneous depolarization is a slow activation of channels permeable to both K^+ and Na^+ (g_f) and a few Ca^{2+} channels (g_{Ca}), the latter brought about by the slow depolarization.

Excitation–contraction coupling is brought about by Ca^{2+} ions

> Every AP raises the Ca^{2+} concentration in the cytoplasm of contractile myocardial cells from $0.1\,\mu M$ to $5\,\mu M$. The Ca^{2+} comes from 2 sources: (1) As in skeletal muscle (p. 82), from the sarcoplasmic reticulum by Ca^{2+}-induced Ca^{2+} release via the ryanodine receptor (SR Ca^{2+} release channel). (2) Also extracellular Ca^{2+} ions flow into the cell during the AP plateau via Ca^{2+} channels (see above) and replenish the intracellular stores. Hence the AP has both a triggering and refilling effect; therefore shortening of the AP (tachycardia) reduces Ca^{2+} uptake and causes a fall in the force of contraction; prolonging the AP has the opposite effect. As in skeletal muscle, contraction is ended by pumping the Ca^{2+} ions back into the sarcoplasmic reticulum. Inhibition of the Ca^{2+} influx by Ca^{2+} antagonists (e.g. verapamil, nifedipine) also weakens the contraction.

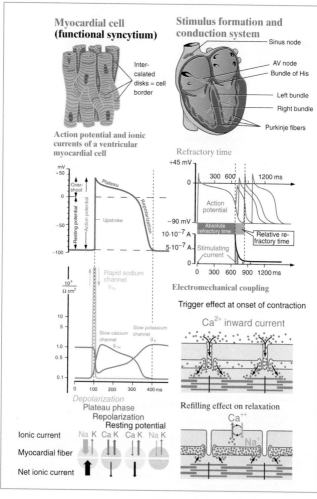

Myocardial cell (functional syncytium)

Inter-calated disks = cell border

Stimulus formation and conduction system

Sinus node
AV node
Bundle of His
Left bundle
Right bundle
Purkinje fibers

Action potential and ionic currents of a ventricular myocardial cell

mV
+50
Over-shoot
Plateau
0
Resting potential
Action potential
Upstroke
Repolarization
−50
−100

Refractory time

+45 mV
0
300   600   1200 ms
Action potential
−90 mV
$10 \cdot 10^{-7}$ A
Absolute refractory time
Relative refractory time
$5 \cdot 10^{-7}$ A
Stimulating current
0   300   600   900   1200 ms

$\frac{10^3}{\Omega \, cm^2}$

Rapid sodium channel
g_{Na}

10
5
Slow calcium channel
g_{Ca}
Slow potassium channel
g_K

1.0
0.5

0.1

0   100   200   300   400 ms

Depolarization
Plateau phase
Repolarization
Resting potential

Electromechanical coupling

Trigger effect at onset of contraction

Ca^{2+} inward current

Ionic current   Na K   Ca K   Ca K   Na K

Myocardial fiber

Net ionic current

Refilling effect on relaxation

Ca^{++}

Na^+

Autonomic and afferent innervation of the heart

The heart receives efferent innervation from parasympathetic fibers of the vagus nerves and from sympathetic nerve fibers (cardiac nerves)

Efferent	Main target site	Main effect	Transmitter/mechanism/comments
Right vagus	SA node	Negative chronotropic (lowers frequency)	Transmitter (also for left vagus): acetylcholine (ACh), flattens pacemaker potential by raising g_K. ACh receptor is muscarinic, blocked by atropine. Slow heartbeat is called bradycardia
Left vagus	AV node	Negative dromotropic (slows conduction)	Conduction in the AV node is slowed by increased g_K. G-protein couples the ACh receptor to the K^+ channels
Sympathetic (right and left)	Whole heart	Sinus node: positive chronotropic Contractile myocardium: positive inotropic	Transmitter: Norepinephrine binds to β_1-adrenoreceptors. Epinephrine (from adrenal medulla) has similar action. Pacemaker potential becomes steeper by activation of Na^+ and Ca^{2+} channels. Rapid heartbeat: tachycardia. Inotropic action also occurs by activation of additional Ca^{2+} channels during plateau of AP

Both efferent systems are tonically active at rest, but in general the vagus tone predominates: blocking of both systems raises the heart rate from ca. 70 to ca. 105 beats/min in a young adult. Exercise raises the sympathetic tone while simultaneously reducing vagus tone, and vice versa.

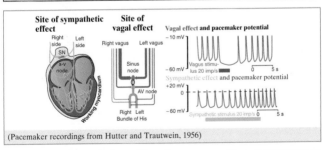

(Pacemaker recordings from Hutter and Trautwein, 1956)

Afferent innervation of the heart: (1) mechanosensors with myelinated fibers at the openings of the large veins into the atria, (2) mechano- and chemosensors with C fibers in the whole heart

Re 1: Afferent fibers in the vagus nerves. Sensors respond in part to atrial contraction (A sensors), in part to filling of the atria, hence stretch (B sensors) or to both (mixed type); all detect degree of atrial filling, induce reflex tachycardia and slight increase in diuresis.
Re 2: Afferent fibers in both the vagus nerves and the sympathetic cardiac nerves. Low-threshold mechanosensors are activated by ventricular contraction and induce bradycardia. Chemosensitive nocisensors mediate angina pectoris and pain of myocardial infarction.

Electrocardiogram (ECG)

> Electrocardiography is the recording from the skin surface of electrical potential differences produced by the depolarization and repolarization of the myocardium; the waveforms recorded constitute the electrocardiogram (ECG)
>
> The spread of excitation within the heart is associated with extracellular currents. These produce small potential differences at the body surface, because the extracellular fluid acts as an electrical conductor between heart and skin. The potential differences (order of magnitude 1 mV) are detected with metal contact electrodes, amplified and traced on a strip of paper.
>
> The size of the potential differences depends on the magnitude of the extracellular currents, and these in turn depend on the number of myocardial cells undergoing a change in membrane voltage. Because the conducting system comprises only a few cells, the ECG reflects only activity of the atrial and ventricular contractile myocardium.
>
> The method was developed around the turn of the century by Willem Einthoven in Leyden and Augustus Waller in London. It allows one to draw conclusions about excitation and conduction, but not about contraction of the heart.

The recording sites and conditions of the ECG are internationally standardized. The most commonly used are Einthoven's bipolar leads:

Limb leads

Connected to:
Right arm
Left arm
Left leg
Right leg   (ground)

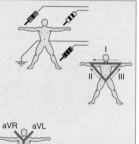

Einthoven's bipolar limb leads:
Lead I   Left arm → Right arm
Lead II   Left leg → Right arm
Lead III   Left leg → Left arm

Chest leads

Wilson's unipolar (precordial) leads:
V_1 4th intercostal, parasternal, right
V_2 4th intercostal, parasternal, left
V_3 Between V_2 and V_4
V_4 5th intercostal in midclavicular line, left (normally apex of heart)
V_5 Anterior axillary line at level of V_4
V_6 Middle axillary line at level of V_4 and V_5

Relationships of the ECG to the sequence of excitatory processes in the heart

Nomenclature: A segment is the distance (time) between two waves, an interval comprises both waves and segments. The RR interval corresponds to the duration of a heart cycle (used to determine heart rate). Positive deflections are shown upward; in the QRS complex the positive deflection is called the R wave.

ECG element	Definition/comments
P wave	Associated with the upstroke of the atrial action potentials, i.e. it signals the spread of excitation in the atria. It is followed by contraction of the atria, during the PR interval. The decay of atrial excitation (repolarization) is not visible in the ECG, due to the small mass and the slow time course of the repolarization.
PQ (or PR) interval, conduction time	Time during which the excitation spreads through the atria (P wave, see above) and passes through the AV node into the bundle of His and the ventricular bundle branches; normal duration 0.18–0.20 s. Most of this time corresponds to transit in the AV node.
QRS complex	Reflects the spread of excitation in the ventricular myocardium. Overall duration ≤ 0.1 s. Depending on recording configuration and heart position, less than three waves may be identifiable
ST segment	Corresponds to the plateau phase of the action potentials in the ventricular myocardium (whole myocardium depolarized, hence no potential differences are recorded); duration highly dependent on frequency (see above)
T wave	Reflects the decay of excitation (repolarization) in the ventricular myocardium; has the same polarity as the R wave, because the path for repolarization is opposite to the path for depolarization. Repolarization is (1) more rapid at the apex than at the base of the heart ('apicobasal recovery') and (2) more rapid near the pericardial than near the endocardial surface of the ventricles

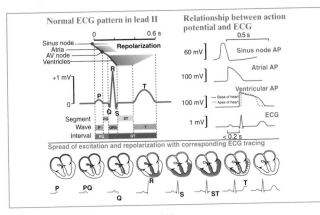

Spread of excitation and repolarization with corresponding ECG tracing

The waves of the ECG can be interpreted as projections of the net electric dipole (called 'integral vector') onto the line joining the recording sites at each moment of the excitation cycle; in vectorcardiography, the curve traced by the dipole tip during the heart cycle is represented

Unexcited parts of the heart have a preserved membrane voltage, i.e. they are outside-positive with respect to excited parts (depolarized cells). Every excited cell acts as an electric dipole with a particular orientation, a dipole vector. The charge, magnitude and direction of each dipole sum with all the other dipole vectors to give the integral vector. This integral vector circles counterclockwise in the frontal plane during ventricular excitation. The position of the tip, relative to time, gives the vector loop.

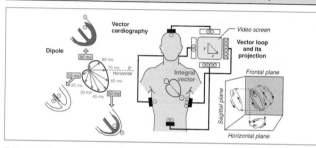

Following the vectorial interpretation of the ECG, the 'electrical position' of the heart in the thorax can be determined using Einthoven's triangle to construct a graph representing the electrical axis of the heart

The direction of the integral vector at the time of the R wave is called the electrical axis of the heart. It can be found graphically from any two standard limb recordings. For simplicity it is assumed that the recording sites form an equilateral triangle, onto the sides of which the electrical heart axis is projected. Much empirical evidence shows that this corresponds quite closely to the anatomical position of the heart in the thorax.

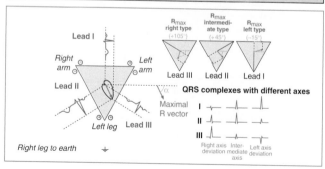

QRS complexes with different axes

The pathological ECG (examples)

Types of extrasystoles:
1. Atrial extrasystole: shift of rhythm

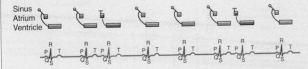

2. Ventricular extrasystole
(a) interpolated

(b) with compensatory pause

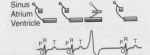

Disorders of propagation of impulse
Complete AV block (3° AV block) with idioventricular rhythm

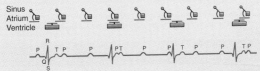

Atrial fibrillation with irregular ventricular rate ('absolute arrhythmia')

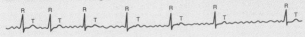

Depression of ST segment: points of infarction/ ischemia

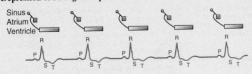

The heart as a pump

Temporal relationships of the various physiological parameters directly linked to activity of the heart (after C. Wiggers, modified by Gregg, 1961 and Gauer, 1972)

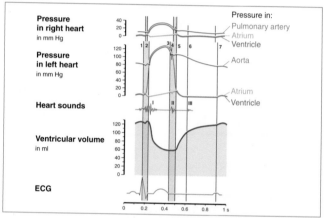

Action phases in the left ventricle (frequency 70/min). The numbers in parentheses show the value in the right ventricle (after C. Wiggers, modified by Gregg, 1961 and Gauer, 1972)

No. in figure	Phase	Time (s)	Comments
	Ventricular systole	**0.27**	
1–2	Isovolumetric contraction	0.06 (0.02)	Isovolumetric increase in tension of ventricle walls. At 2, the aortic value opens
2–4	Ejection period	0.21	Initially fast (rapid ejection), followed by a longer phase of slower flow (reduced ejection). Throughout there are parallel pressure changes in ventricle and aorta. End of systole is not sharply definable
	Ventricular diastole	**0.56**	
3–4	Protodiastole	0.02	Brief transition period, with loss of fiber tension
4–5	Isovolumetric relaxation	0.05 (0.02)	Begins with closure of the aortic valve
5–6	Rapid filling period	(0.16)	At 5 the ventricular pressure falls below the atrial pressure, the mitral valve opens, the ventricle fills to 80% capacity
6–7	Diastasis	0.23	Slow filling of the remaining 20%
7–1	Atrial systole	0.10	Contribution to filling is small at normal heart rate, but important if there is tachycardia

Influence of frequency on cardiac mechanics

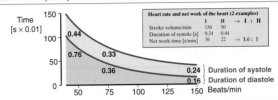

Heart rate and net work of the heart (2 examples)			
	I	II	→ I : II
Stroke volume/min	150	50	
Duration of systole [s]	0.24	0.44	
Net work time [s/min]	36	22	→ 1.6 : 1

Increased frequency occurs primarily by decreased duration of diastole. At 90–100 beats/min diastole and systole are about equally long; at higher rates, diastole becomes gradually shorter. From 150 beats/min on, ventricular filling is reduced, causing a decreased stroke volume. At very high frequencies ('ventricular flutter') there is no filling of the ventricles.

Heart valves (located in the connective tissue plate between atrium and ventricle, called the valve plane)

Like valves in a piston pump, they function to direct flow. We distinguish them according to construction and location:
1. Cusp valves (a) Mitral valve = Atrioventricular valve (AV valve), left
              (b) Tricuspid valve = AV valve, right
2. Semilunar valves (a) Aortic valve at aortic ostium
              (b) Pulmonary valve at pulmonary ostium

Heart valve defects: Valves are malformed or altered by chronic inflammation. Stenosis (high resistance to flow) or insufficiency (inability to close completely) in single valves or in combination. Myocardial work must increase to maintain cardiac output. Normal major heart sounds are produced by closure of valves (see below).

Valve plane mechanism

During each ventricular systole a large fraction of the stroke volume of the next systole is sucked into the atria by a shift of the valve plane toward the apex (valve plane mechanism). Because of it, the duration of diastole is less important for ventricular filling than would be expected. At the end of systole the atria are full to capacity. When the AV valves open, atrial blood surges into the ventricular cavity, while the valve plane moves away from the apex, as though the ventricle were being pulled over the column of fluid.

Signals of heart activity recordable at the body surface (without ECG)

Term	Signal of . . . / comments
Apex impulse	Beginning of mechanical systole. Each contraction causes impact in the left 5th intercostal space
Heart sounds	Beginning and end of systole; no delay. For details see below
Venous pulse	Atrial systole, end of valve plane movement and of rapid atrial filling. Delayed with respect to causative events
Arterial pulse	Beginning of ejection period, sometimes end-systolic valve closure; delayed with respect to causative events

Heart sounds and murmurs, recorded by phonocardiography

Event	Definition/comments
1st heart sound	Mainly produced by closure of mitral valve. Also isometric contraction of the ventricles, i.e. isovolumetric contraction period (phase 1–2 in Fig., p. 166)
2nd heart sound	Occurs at closure of semilunar valves (phase 4)
3rd heart sound	Facultative sound, denotes sudden termination of rapid filling period (phase 6)
Heart murmurs	Caused primarily by turbulence at stenotic or insufficient valves. Timing relative to contraction cycle (systolic and diastolic murmurs), duration and character of the sound are important diagnostic criteria

Laplace's law: relation between wall tension T and internal pressure p in sphere (approximates left ventricle) and cylinder (approximates blood vessels)

	Sphere	Cylinder (lateral stretch)
$\text{Tension} = \dfrac{\text{Force}}{\text{Cross-section}}$	$T = \dfrac{r}{2d}\, p$	$T = \dfrac{rp}{d}$

r, radius of sphere or cylinder; d, wall thickness. In both cases the force T that resists disruption of the wall is given by the product of cavity cross-section times pressure, e.g. $\pi r^2 p$ for the sphere and $2rlp$ for the cylinder (where l is length of cylinder). In general terms, the cross-sectional tension is proportional to the product of pressure and radius: $T \propto pr$. Hence, for the same pressure, the tension decreases as the radius decreases.

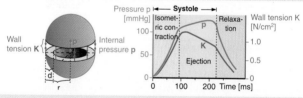

Summary for heart. (1) During systole, the tension of the myocardial fibers decreases as blood is ejected, although the pressure remains the same or shows a moderate increase; the reason is that the ventricle radius decreases and the thickness of the ventricle wall increases. (2) Small hearts can generate high pressures by developing a relatively smaller force per single muscle fiber.

Summary for vessels. In small-diameter vessels, the wall tension is small even with high internal pressures. Arterioles and capillaries can therefore easily tolerate high pressures (e.g. 100 mmHg).

Pathophysiology note. When a wall is overstretched (e.g. heart dilation, aortic aneurysm) a vicious circle (positive feedback) develops, because any additional stretching causes the wall tension (with constant internal pressure!) to rise continually, eventually resulting in cardiomegaly or rupture of aortic wall, respectively.

The elementary forms of contraction of heart versus skeletal muscle

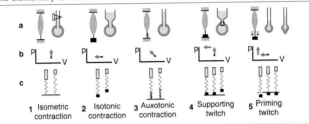

1 Isometric contraction **2** Isotonic contraction **3** Auxotonic contraction **4** Supporting twitch **5** Priming twitch

a Mechanical conditions; **b** pressure–volume diagram (or length–tension diagram); **c** behavior of elements with passive elastic and with contractile function. Example 4 is closest to the normal situation. The weight is supported by a table; the hydrostatic pressure of the fluid bears on the valves. During contraction, tension rises until the weight is just barely lifted or the valves open. Then tension remains constant during shortening or pressure constant during volume flow, respectively. The (diastolic) fiber tension caused by passive filling of the heart is called preload; the (systolic) fiber tension at the moment of valve opening, afterload. According to Laplace's law, both depend on ventricular pressure, radius and wall thickness; the afterload is also determined by end-diastolic pressure (see Frank–Starling law).

Control of stroke volume by the heart itself: Frank–Starling mechanism (law)

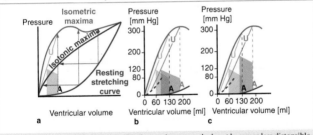

a Work diagram of the isolated heart. Passive tension curve: the heart becomes less distensible, the greater its filling; the isovolumetric (closed aortic valve) or isotonic contractions elicited in various states of filling give the maximal attainable pressure and volume, respectively and are joined to give the curves of isovolumetric or isotonic maxima. With afterloaded contraction (first isovolumetric tension, then shortening/ejection) the maxima cannot be reached (curve A, see also b, c). **b** Dependence of stroke volume on end-diastolic filling. End-diastolic 130 ml: shaded area, end-diastolic 180 ml: red area (stroke volume nearly doubled, negligible increase in end-diastolic volume). **c** Work-diagram shift due to increased aortic pressure. With initially reduced stroke volume (not shown) a certain end-systolic volume remains and is added to the normal venous inflow (working point W shifts to the right). The result is an increased stroke volume, until with greatly increased end-diastolic and end-systolic volumes the original stroke volume is ejected against the elevated aortic pressure.

Influence of inotropy on cardiac work

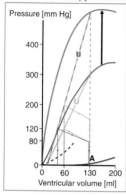

Sympathetic activity (or epinephrine/norepinephrine) shifts the curve of isovolumetric maxima upward (red arrow). The curve of afterloaded contraction becomes correspondingly steeper. In other words: for a given end-diastolic volume, sympathetic stimulation will cause the heart to eject either (1) the same stroke volume against higher pressure or (2) a larger stroke volume against the same pressure. Such changes in inotropy elicited by changes in sympathetic tone are the normal form of adaptation of the heart to changing loads. The Frank–Starling mechanism is also important, e.g. for adjusting the amounts ejected by the two ventricles, preventing large changes in pulmonary blood volume, and in light-to-moderate exercise.

Contraction velocity and contractility (syn.: inotropy)

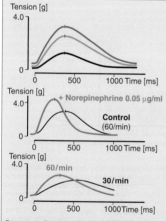

Isometric contraction of a cat papillary muscle with increasing passive tension: maximal tension and contraction velocity increase, but the peak always occurs at the same time.

Norepinephrine increases isometric tension with a simultaneous rise in velocities of contraction and relaxation.

Increase in contraction frequency from 30/min to 60/min raises the tension slightly and the contraction velocity considerably (after Sonnenblick, 1962).

Summary: Increased contractile force by norepinephrine is accompanied by a parallel increase in contraction velocity. Contractility (inotropy) can therefore be well evaluated in humans by measuring the maximal velocity of pressure increase (dP/dt) in the isovolumetric tension phase (by a catheter in the left ventricle; normal values 1500–2000 mmHg/s = 200–333 kPa/s).

Normal heart size, hypertrophy due to endurance training (modified after Reindell *et al.*, 1967)

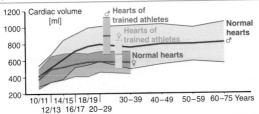

Dependence of heart size on sex and age (means and range of variation while recumbent, based on over 400 measurements). On the right, heart sizes of high-performance athletes. By subtracting from total heart volume (800 ml in man) the volume of the myocardium (corresponds roughly to the heart weight, 300 g), one obtains the volume of blood in the cardiac cavities. Endurance training increases size (both length and thickness) of the individual muscle fibers (not their number).

Measurement of cardiac output (volume per unit time) by Fick's principle

Cardiac output is the product of stroke volume (ca. 70 ml) and heart rate (ca. 70/min), i.e. about 5 liters per min. In 1870 Adolf Fick realized that cardiac output is quantitatively related to O_2 uptake in the lung and can be calculated from pulmonary O_2 and the O_2 concentration difference between arterial blood and venous mixed blood (heart catheter in right ventricle or pulmonary artery), e.g. at rest: O_2 uptake 250 ml/min, O_2 concentration: arterial, 200 ml/l, venous mixed blood, 150 ml/l; hence every liter of blood has taken up 50 ml O_2 in flowing through the lung, so that the amount flowing through must have been 5 l/min. That is:

$$\text{Cardiac output} = \frac{O_2 \text{ uptake}}{O_2 \text{ concentration difference}} = \frac{250 \text{ ml/min}}{50 \text{ ml}} = 5 \text{ l/min}$$

Circulatory variables at rest and during hard muscular work (ca. 12 400 J/min). Fit but not specially trained subject (after Asmussen and Nielsen, 1955)

	Resting value	Working value	Work:rest ratio
O_2 consumption (ml/min)	250	3 000	12.0 : 1
Cardiac output (ml/min)	5000	19 000	3.8 : 1
Arteriovenous O_2 difference (ml/l blood)	45	140	3.1 : 1
Stroke volume (ml)	91	128	1.4 : 1
Pulse rate (per min)	64	174	2.7 : 1

Summary: The increase in O_2 consumption to 12-fold is accomplished by increasing the cardiac output 3.8-fold and raising the amount of O_2 withdrawn from the blood 3.1-fold. Because the heart is so constructed that the stroke volume can be increased only just under 1.5-fold, the elevated cardiac output must be achieved primarily by raising the heart rate.

Energetics of the heart

Calculation of cardiac work and power

The heart as a pump performs primarily pressure–volume work, by displacing a volume (V) under pressure (p) against a resistance to flow. In addition, there is a normally small acceleration work, by which the stroke volume with mass m (2×70 ml) is accelerated to the ejection velocity v (0.5 m/s).

Pressure–volume work:	$p \times V$ (resting values)	
Left ventricle p = 100 mmHg V = 70 ml	$= 100 \times 133$ N/m^2 $= 70 \times 10^{-6}$ m^3	$p \times V = 0.932$ N m
Right ventricle p = 15 mmHg V = 70 ml	$= 15 \times 133$ N/m^2 $= 70 \times 10^{-6}$ m^3	$p \times V = 0.140$ N m
Acceleration work: $\frac{1}{2}$ mv^2	m = 2×70 g v = 0.5 m/s	$2 \times \frac{1}{2}$ mv$^2 = 0.018$ N m
Total work at rest:		Work = 1.089 N m

Cardiac power and power-to-weight ratio

Cardiac power (work per unit time) at the frequency 1 systole/s (60/min) is 1 N m/s = 1 W (0.1 kgf·m/s). Power-to-weight-ratio for a heart weighing 3 N (300 g) is thus 3 N/W = 3000 N/kW. Automobile motors perform considerably better, with only 40–70 N/kW. During exercise the cardiac power rises and the power-to-weight ratio approaches those for man-made pumps.

O$_2$ consumption (see Table p. 171) and efficiency of the heart musculature

At rest the heart consumes 24–30 ml O$_2$/min, i.e. 10% of the total uptake, although it weighs only 300 g (0.5% of the total body weight). During exercise O$_2$ consumption increases several fold (Table). The efficiency (fraction of the total energy expenditure that is converted into mechanical energy) at rest is of the order of 15–20%. It depends on the nature of cardiac work: volume work (relatively high venous input) is more efficient than pressure work (relatively high peripheral resistance). The efficiency increases with exercise; the heart of a trained individual is more efficient than that of an untrained subject.

Substrates of oxidative cardiac metabolism at rest and during physical work
(after Keul *et al.*, 1965)

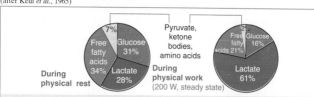

Summary: The myocardium uses multiple metabolic fuels: one major substrate is lactate, which is abundantly available during muscle work (but is not metabolized by skeletal muscle).

Coronary blood flow

The heart is supplied with blood by the right and left coronary arteries
(after Töndury, 1969)

| 50% of all hearts | 30% of all hearts | 20% of all hearts |

Normal values for myocardial flow rate in humans

| Resting value | 80 ml/100 g/min (hence total of 240 ml/min, ca. 5% of cardiac output) |
| Maximal work | Increase to 4- or 5-fold (called coronary reserve) |

Variation in coronary blood flow during the heart cycle (after Antoni, 1989)

The small vessels in the heart pass through the middle and inner layers of the wall of the ventricles, where they are subject to the effects of myocardial contraction. As a result, in the left ventricle at the beginning of systole, the influx of blood is almost completely suppressed and the maximum is not reached until relaxation begins. In contrast, in the right ventricle, isovolumetric contraction is weaker and hence coronary flow persists. During systole, venous blood is pressed out of the coronary sinus.

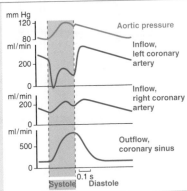

Regulation of coronary blood flow

Regulation is brought about mainly by local chemical (metabolic) factors: vessels are dilated by a fall in O_2 partial pressure, but also by adenosine (product of the breakdown of energy-rich phosphates), by an increase in the extracellular K^+ concentration and by factors of endothelial origin. In addition, sympathetic activity has two kinds of effects on coronary blood flow: (a) indirect, mediated by the inotropic effect (see effect of myocardial contraction, above), and (b) direct, vasoconstricting action. The vagus (parasympathetic) exerts a dilating action.

Survey of arterial and venous circulation

Cardiovascular system in outline, and the distribution of cardiac output to the circulatory subdivisions at rest (after Witzleb, 1989, and Rein and Schneider, 1971)

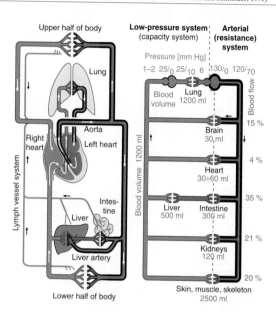

Left: Subdivisions of the circulatory system, diagrammatic. Sections with O_2-saturated 'arterial' blood drawn in light red, those with partially unsaturated 'venous' blood in dark red. Lymph vessels dark gray. *Right:* Percentage of cardiac output allotted to individual organs at rest (cardiac output ca. 5–6 l/min). More detailed data in the following tables and figures.

Summary: The pulmonary circulation with right ventricle as pump is the minor subdivision, in series with the systemic (organ) circulation (major subdivision), with the left ventricle as pump. The arterial system, endowed with a high resistance to flow, extends from the left ventricle to the capillary system; the rest of the circulatory system has a high compliance at low pressure and constitutes the low-pressure or capacitance system. The pulmonary system, situated between right and left heart, also belongs to the low-pressure system.

Distribution of pressures, volumes, vessel cross-sections and flow velocities in the blood circulatory system (modified after Gauer, 1972, and Witzleb, 1989)

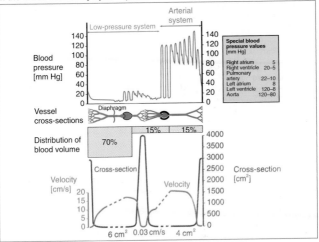

Distribution of blood volumes (in ml) among the subdivisions of the circulatory system in humans. (Averages for ca. 40-year-old, 75 kg man [after Milnor, 1974])

Region	ml	ml	%	%
Heart (diastole)		360	7.2	7.2
Pulmonary circulation				
Arteries	130		2.6	
Capillaries	110	440	2.2	8.8
Veins	200		4.0	
Systemic circulation				
Aorta, large arteries	300		6.0	14
Small arteries	400		8.0	
Capillaries	300	4200	6.0	84.0
Small veins	2300		46.0	64
Large veins	900		18.0	
Total:		5000		100.0

Summary: Most of the blood is in the systemic circulation (84%), and mostly in the veins (3200 out of 4200 ml). In systemic and pulmonary veins, about 70% of the blood is stored (veins are capacitance vessels). Capillaries, despite their large total cross-sectional area, are so short that they contain relatively little blood (410 ml altogether). The small arteries also contain little blood, especially the arterioles, which present the main resistance to flow in the circulatory system (arterioles are resistance vessels).

Mean flow velocities and mean pressures in the human systemic circulation
(after Witzleb, 1989)

	Diameter (mm)	Mean velocity (cm/s)	Mean pressure (mmHg)
Aorta	20–25	20	100
Medium-sized arteries		10–5	95
Very small arteries		2	70–80
Arterioles	0.06–0.02	0.3–0.2	35–70
Capillaries			
Arterial end			30–35
Middle	0.006	0.03	20–25
Venous end			15–20
Very small veins		0.5–1.0	10–15
Small to medium veins		1–5	10
Large veins	5–15	5–10	or
Venae cavae	30–35	10–16	less

Volume flow, resistance to flow (R) and O_2 uptake in individual human organs under resting conditions. Body weight 70 kg, body surface area 1.7 m²
(after Wade and Bishop, 1962 and Witzleb, 1989)

Vascular bed in:	Blood flow ml/min	%	ml/s	R Pa ml⁻¹ s	O_2 uptake ml/min	%	Weight g	%
Splanchnic	1400	24	23	580	58	25	2 800	4.0
Kidneys	1100	19	18	740	16	7	300	0.4
Brain	750	13	13	1025	46	20	1 500	2.0
Heart (coronary)	250	4	14	3330	27	11	300	0.4
Skeletal muscle	1200	21	20	670	70	30	30 000	43.0
Skin	500	9	8	1670	5	2	5 000	7.0
Other organs	600	10	10	1330	12	5	30 100	43.2
Total systemic circulation	5800	100	96	140	234	100	70 000	100.0
Pulmonary system	5800	100	96	11				

Total blood flow, specific blood flow and arteriovenous O_2 difference (AVO$_2$ in human organs under resting conditions and with maximal vasodilation (after Golenhofen, 1981; some resting blood flow values differ slightly from those in above Table and Fig. p. 175)

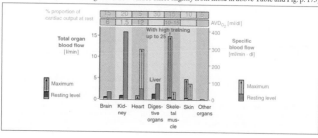

Hemodynamics

Flow in rigid tubes: dependence on pressure and resistance (Fig. based on Max von Frey, modified from Rein and Schneider, 1971)

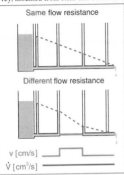

Same flow resistance

Different flow resistance

v [cm/s]

$\dot{V}$ [cm³/s]

Liquid flow in a tube is driven by the pressure difference ($\Delta p = p_1 - p_2$) between its ends. This pressure difference provides the force needed to overcome resistance to flow. If the latter varies, so does the pressure drop, as can be seen by the levels to which the liquid rises in the vertical tubes at different positions. The volume flow ($\dot{V}$, volume flowing per unit time), the pressure gradient Δp and the resistance to flow R are related by a simple rule analogous to Ohm's law:

$$\dot{V} = \frac{\Delta p}{R} \qquad (1)$$

During steady-state flow, $\dot{V}$ is the same in all sections of the tube (law of continuity); hence the linear flow velocity v (velocity of an infinitesimal volume element) must change in inverse proportion to the cross-sectional areas of the individual sections.

Resistance to flow R and volume flow $\dot{V}$ in a rigid thin tube: Hagen–Poiseuille law

R depends on the 'friction' of the flowing material. This is determined by the fluid viscosity η and the dimensions of the tube: length L and radius r. With respect to the latter, it has been found that the energy loss due to friction is inversely proportional to the 4th power of the tube radius. In summary, resistance, R is given by

$$R = \frac{8\eta L}{\pi r^4} \qquad (2)$$

When this value is substituted in equation (1), $\dot{V}$ is given by

$$\dot{V} = \frac{\pi r^4 \Delta p}{8\eta L} \quad \text{(Hagen–Poiseuille)} \qquad (3)$$

Summary: Very small changes in tube diameter greatly alter the resistance to flow. Therefore changes in r are considerably more effective than other factors to bring about a change in the resistance.

Comments on the Hagen–Poiseuille law

The law applies only to laminar flow (flow in layers), but this is practically always the case in the circulatory system (axial flow is faster than peripheral flow due to friction at the wall). Turbulent flow is about equally rapid, but has a higher resistance due to its internal vortices. From a critical velocity on, laminar flow can become turbulent (affected by several factors, including tube diameter and viscosity). The limiting value is expressed as the critical Reynolds' number (Re_c). The viscosity η of human blood is 3–5 relative units, and for plasma is 1.9–2.3 relative units. These values hold only for relatively rapid flow and normal blood composition. During slow flow η increases, in part because the erythrocytes form 'rouleaux' (pathologically important, e.g. in shock: flow falls because R is too high). Conversely, effective η decreases in narrow vessels (<500 μm): Fåhraeus–Lindqvist effect (plasma forms a low-friction peripheral layer, with cells confined to the axial current).

Flow in an elastic vessel: pressure, vessel diameter, wall tension, elasticity

The walls of vessels have considerable elasticity. Therefore, as pressure increases the vessels are progressively stretched, so that the resistance to flow falls and the volume flow rises with a slope greater than linear. The stretching (transmural) pressure p_t is the pressure difference between inner surface and outer surface of the vessel wall ($p_t = p_i - p_o$ [where p_o is usually very small, so that $p_t \sim p_i$]). p_t generates a tangential tension T in the vessel walls that opposes the pressure; according to Laplace's law (see p. 168). T also depends on the inside radius r_i and the wall thickness h. This is illustrated in the following figure and table.

Transmural pressure p and tangential wall tension T in various vessels with inside radius r and wall thickness h (from Witzleb, 1989 based on data of Burton, 1969); Re p and r cf. Table p. 176

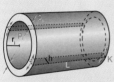

Vessels	r (mm)	h (mm)	p (mmHg)	T (N/m)
Aorta	12	2	100	173
Arteries	0.5–3	0.3	90	24
Arterioles	0.01–0.1	0.03	60	0.48
Capillaries	0.003	0.001	30	0.016
Venules	0.01–0.025	0.002	20	0.027
Veins	0.75–7.5	0.5	15	1
Vena cava	17	1.5	10	21

Summary: The decrease in tangential wall tension (tensile stress) in thin vessels is far greater than the decrease in pressure, as explained by Laplace's law. Therefore even the thin walls of the arterioles and capillaries can easily tolerate the blood pressure there, which is still relatively high. If the vessel wall were split longitudinally along L in the figure, the cut edges would spring apart with the force T.

Autoregulation of the vessels can increase the degree of contraction of their walls and thereby counteract elastic vessel dilation

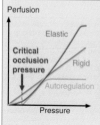

The smooth muscle of the vessel walls has a certain automatism (ability to contract or raise tone), which can be activated or enhanced by stretch. When the transmural pressure increases, this effect (Bayliss effect) can be so strong that further stretch is counteracted by progressive contraction. Accordingly, despite the greater pressure there is no substantial increase in blood flow. This behavior is typical, e.g. of arteries in kidneys and brain. Furthermore, it should be kept in mind that the capillary bed cannot be opened until the pressure is above a certain minimum, about 20 mmHg, the critical closing pressure (found by extrapolating the steep part of the curve to the abscissa). This closing pressure reduces the effective pressure gradient.

The 'Windkessel' function of the arterial system

By incorporating an air-filled chamber (a 'Windkessel', left) into a fire hose, discontinuous flow (e.g. flow driven by an intermittent pump) can become more continuous: after the pump valve has closed, the energy stored in the chamber keeps the flow going during the pump 'diastole'. In the vascular system, right, the elastic restoring forces of the stretched vessel walls take over the role of the air cushion. The variable line density symbolizes that the blood vessels are more readily stretched near the heart (aorta, see below) and much less so toward the periphery. Nevertheless, blood flow becomes continuous only in the microcirculation.

Pressure–volume (pV) diagram of the human aorta, elasticity and total capacity of the arterial Windkessel (values taken from Simon and Meyer, 1958)

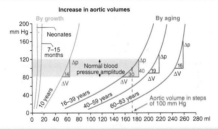

The volume and elasticity of the aorta change with growth and aging. The curves for all aortae are concave upward, i.e. the compliance decreases at higher pressures. The slope of the pV curve, given by $E' = \Delta p/\Delta V$, quantifies the Windkessel effect. Greatest elasticity (see top of preceding page) occurs in early adulthood. Average capacity of the aorta at $100\,\text{mmHg}$ is $170\,\text{ml}$, or about 20% of the total capacity of the arterial system, $850\,\text{ml}$ (15% of the blood volume). E' of the aorta can be found from the figure as

$$E' = \frac{\Delta p}{\Delta V} = \frac{40\,\text{mmHg}}{30\,\text{ml}} = 1.33\,\text{mmHg per } 1\,\text{ml} = 177\,\text{Pa/ml}$$

The aorta is about three times more distensible than the rest of the arterial system, which accepts only $10\,\text{ml}$ for a pressure increase of $40\,\text{mmHg}$ (the normal blood pressure amplitude). The overall Windkessel effect therefore has an E'

$$E_{overall} = \frac{40\,\text{mmHg}}{30\,\text{ml} + 10\,\text{ml}} = 1\,\text{mmHg per } 1\,\text{ml} \left(= 133\,\text{Pa/ml}\right)$$

Summary: An increase in arterial pressure by $1\,\text{mmHg}$ causes the arterial volume to increase by ca. $1\,\text{ml}$ and vice versa. In old age the elements of the aortic wall become less distensible, and therefore the mean pressure increases. Both factors make it necessary for the aging heart to work harder.

The arterial pulse

Shape of the pressure pulse and its velocity in various arteries (measurements by O.H. Gauer with intra-arterial catheter, beginning in left ventricle)

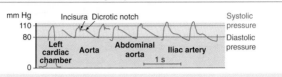

Closure of the aortic valves is reflected in the incisura. The elevation of the pulse curve in diastole (dicrotic wave) and the overshooting pulse amplitude in the periphery are probably caused by reflections and the development of standing waves in the arteries. Blood flow is crucially determined by the arterial mean pressure, values of which are given in the Table p. 176. The pulse-wave velocity in the aorta is 4–6 m/s; it increases toward the periphery, because the vessel walls become more rigid (e.g. radial artery 10–12 m/s). Therefore the pulse-wave velocity also rises with increasing arterial mean pressure and in old age (in both cases due to more rigid vessel walls, see figure on p. 179).

Qualities of the pulse (palpable with fingertips, measurable with pressure transducers or intra-arterial catheters)

Quality	Terms	Definition/comments
1. Frequency	Tachycardia Bradycardia	Simplest method of measuring heart rate. Normally 60–80/min, higher in children, lower in trained athletes
2. Regularity	Regular pulse Irregular pulse	Physiological: respiratory (sinus) arrhythmia: increase in frequency during inspiration, especially in juveniles
3. Tension	Hard pulse Soft pulse	Rough estimation of systolic pressure by testing the force required to suppress the pulse (pressing the artery against a hard surface)
4. Amplitude	Strong pulse Weak pulse	Blood pressure amplitude permits rough estimation of stroke volume, the basic determinant of pressure amplitude with other factors constant. Strong pulse in old age due to greater arterial wall rigidity
5. Sharpness	Short pulse Long pulse	Strong pulse is usually also short, because the pressure changes more rapidly per unit time. Similarly, weak pulse is usually long

Volume pulse, current pulse, flow velocity

As the stroke volume is ejected, the wall of the aorta is stretched due to its elasticity (cross-section increases), increasing its volume by about 30 ml (see figure on preceding page). In the course of diastole the aortic cross-section decreases (Windkessel effect). The time course of this aortic volume change is called cross-sectional or volume pulse. It corresponds closely to the pressure changes in the aorta. Ejection of the stroke volume also causes flow of blood in the aorta. This current pulse has a peak velocity of about 150 cm/s (average during ejection 70 cm/s). After the end of ejection the column of blood in the ascending aorta practically stops moving or may even briefly reverse. Therefore, the average flow velocity over the entire cycle is only 20 cm/s (see p. 176 and the figure on p. 175; the figure also shows velocities for the rest of the arterial system).

Blood pressure and its measurement

Nomenclature, normal values

Term	Definition/comments
Systolic pressure	Maximal arterial pressure at the peak of the ejection period (see figures on pp. 166, 175). Normal average in recumbent young adults in aorta and large arteries 120 mmHg
Diastolic pressure	Lowest arterial pressure, occurs at the end of the isovolumetric contraction period (i.e. during systole), before the aortic valves open. Normal average in aorta and large arteries 80 mmHg
Pulse pressure	Difference between systolic and diastolic pressures. Normal average in aorta and large arteries 40 mmHg
Mean pressure	Calculated from the integral of the pressure over time, is approximately equal to the diastolic pressure plus one-third of the pulse pressure, ca. 90–95 mmHg. For other mean pressures in vascular system see p. 176

Indirect blood pressure measurement, methods of Riva-Rocci and Korotkoff

An inflatable cuff (standard width 12 cm, narrower for children) connected to a pressure gauge is wrapped around the upper arm. Riva-Rocci method: pulse is felt at the radial artery. When cuff pressure exceeds systolic blood pressure, pulse disappears; cuff pressure is then reduced, and when it falls below systolic pressure pulse reappears. Diastolic pressure cannot be measured in this way. Korotkoff method detects both systolic and diastolic pressures. Cuff inflated to a pressure above systolic blood pressure and slowly deflated; as soon as cuff pressure falls below systolic blood pressure, the blood flow through the compressed artery is turbulent. Turbulence causes sharp Korotkoff sounds, which can be heard with a stethoscope over the cubital artery and indicate systolic value. With further reduction of cuff pressure, sound first becomes louder and then rapidly declines. As soon as cuff pressure falls below diastolic pressure, the Korotkoff sounds disappear, because the artery is completely open. Hence sound disappearance signals diastolic pressure.

Age dependence of human blood pressure (data from US National Health Survey 1960–1962, comprehensive statistical sample from entire white population)

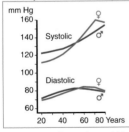

Upper limits of normal (rounded from National High Blood Pressure Education Program Coordinating Committee, 1990)

Systolic blood pressure (rule of thumb: 100 + age):

140 mmHg up to 40th year of life,
150 mmHg from 40th to 60th year,
160 mmHg from 60th year on.

Diastolic blood pressure:

80 mmHg up to 40th year of life,
90 mmHg from 40th to 60th year,
90 mmHg from 60th year on.

Higher-order fluctuations of blood pressure

The pulse is a first-order fluctuation. Those of second order are synchronous with breathing: at each inspiration the blood pressure decreases slightly and the pulse is accelerated; vice versa during expiration (central nervous coupling of breathing and circulation, also mechanical component). Hering–Traube–Meyer waves, with periods of 10s or longer, are fluctuations of third order (slow fluctuations of vascular tone about a mean). Circadian rhythm causes 24 h fluctuations of fourth order with a minimum in the early morning hours.

Venous system (low-pressure system)

The static blood pressure (mean filling pressure) is a measure of the state of filling of the entire vascular system (Fig. modified from Broemser, 1938)

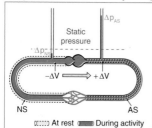

After cardiac arrest the pressure within the cardiovascular system falls to an equilibrium value. This static blood pressure ('mean circulatory filling pressure' for the entire system, and 'mean systemic filling pressure' for the systemic circulation) reflects the relationship between capacity and blood volume. The mean systemic filling pressure is 6–8 mmHg. The heart pumps an amount of blood ΔV out of the low-pressure system LS (including pulmonary circulation) into the arterial system AS. Therefore pressure in the AS rises to the level of the arterial blood pressure, and pressure in the LS falls to that of the central venous pressure (pressure in right atrium).

Because the two systems differ greatly in elasticity, the marked increase in arterial pressure (Δp_{AS} to 120/80 mmHg) is accompanied by only a slight drop in central venous pressure Δp_{LS}, from 6 to 2–4 mmHg. Nevertheless, by establishing the pressure gradient for venous return the latter determines the magnitude of the stroke volume (for venous pressures and flow velocities see Tables pp. 176, 178).

Important: The activity of the heart cannot raise the static blood pressure (mean filling pressure) but can only make its components unequal.

Mechanisms contributing to venous return

In addition to the central venous pressure, several factors contribute to venous return. The main ones are:

Muscle–vein pump	Compression of the muscle veins during muscle contraction. The valves in the veins prevent backflow. (See figure on p. 190)
Suction pressure pump effect of respiration	Negative intrathoracic pressure during inspiration exerts suction on adjacent vessels and lowers vascular resistance. At the same time the pressure in the abdominal cavity rises and presses venous blood toward the heart
Valve plane mechanism	See p. 167

Venous pulse in jugular vein, in temporal relation to heart sounds and ECG
(modified after Wiggers, 1954)

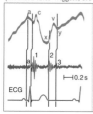

Venous pulse originates in the large veins near the heart, by retrograde pressure transmission during heart contraction. Propagation velocity ca. 1.5 m/s. The following waves can be distinguished:

a-wave:	atrial contraction
c-wave:	bulging of the AV valve
x-depression:	shift of the valve plane towards the apex
v-wave:	pressure rise before AV valve opens
y-depression:	fall after AV valve opens, followed by sharp rise to a-wave

Influence of hydrostatic pressure on arterial and venous pressures when standing erect; indifference level (modified after Guyton, 1976, from Witzleb, 1989)

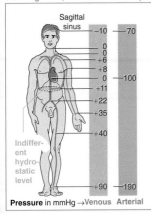

Gravity creates hydrostatic pressures in the circulatory system, which are negligible when lying down and greater under orthostatic conditions (standing upright). Arteries and veins are equally affected, so that the pressure difference between them does not change. The transmural pressure increases, however, so that during standing 400–600 ml blood 'is pooled' in the veins of the lower extremities.

Hence, the pressure is lower in the head, and higher in the feet. The hydrostatic indifference level, at which the pressure does not change when position changes, is 5–10 cm below the diaphragm. Above this level, the pressure is lower when standing than when lying down. In the thorax the subatmospheric pressure prevents collapse of the veins.

Changes in the circulation parameters elicited by changes in position (orthostatic regulation) (modified after Bevegard *et al.*, 1960)

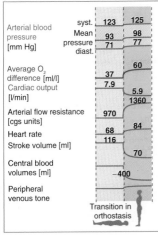

		Venous	Arterial
Arterial blood pressure [mm Hg]	syst.	123	125
	Mean pressure	93	98
	diast.	71	77
Average O₂ difference [ml/l]		37	60
Cardiac output [l/min]		7.9	5.9
Arterial flow resistance [cgs units]		970	1360
Heart rate		68	84
Stroke volume [ml]		116	70
Central blood volumes [ml]			−400
Peripheral venous tone			

Transition in orthostasis

When a person stands up, blood is pooled in the lower extremities because the veins do not contract in response to the increase in hydrostatic pressure (see above). The blood is drawn mainly from the intrathoracic volume, so that heart filling and stroke volume decrease (Frank–Starling, p. 169). By the baroreceptor reflex, this initiates peripheral vasoconstriction (increased resistance) and an increase in heart rate. But the latter does not fully compensate for the decrease in stroke volume and thus cardiac output decreases. Because of the vasoconstriction, there is no fall in blood pressure.

If too much blood pools during standing, the heart is no longer adequately filled, cardiac output falls, and the insufficient brain blood flow leads to orthostatic collapse (which is self-limited, i.e. circulatory parameters recover once the person has fallen down).

Microcirculation and the lymphatic system

Exchange processes and the driving forces in the terminal vascular bed

Process	'Forces'	Description/comments
Diffusion	Concentration difference	(a) Water and water-soluble substances diffuse through membrane pores, lipid-soluble substances (O_2, CO_2, alcohol) through the whole capillary membrane; diffusion is the most important process for exchange of water and solutes in the capillaries. Water is exchanged between plasma and interstitial space ca. 40 times per capillary passage. Daily values in figure on p. 185
		(b) Passage of large molecules (protein) through a few very large pores occurs only to a limited extent (and with great regional differences; liver capillaries have highest permeability). Protein permeating the capillary wall is carried away by lymph
Bulk flow	Hydrostatic and colloid osmotic pressures	The effective filtration pressure is calculated as the algebraic sum (preserving the signs) of the hydrostatic pressures in the capillaries p_c (blood pressure) and of the interstitial fluid p_{IF} and the corresponding colloid osmotic pressures Π_c and Π_{IF}. In a hypothetical 'average' capillary, p_c is high at the beginning and low at the end (values in Table p. 176). Hence, filtration would occur at the beginning and reabsorption at the end (see below)

Structure of terminal vascular bed and net fluid movement across capillary membrane (modified from Gauer, 1972 and Witzleb, 1989, from data of many authors)

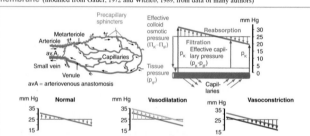

The rate of filtration of fluid ($\dot{V}$) is related to the filtration coefficient (K_f), according to:

$$\dot{V} = K_f\left(p_c - \Pi_c + \Pi_{IF} - p_{IF}\right)$$

$\dot{V}$ is positive for outward filtration, negative for inward filtration (reabsorption). Π_c is 25 mmHg, Π_{IF} is 5 mmHg, p_{IF} is 3 mmHg. Vasodilation or constriction (of arterioles) raises or reduces p_c, respectively. On average 90% of the filtered fluid is reabsorbed, and 10% is removed in the lymphatic system

Total number of capillaries, exchange surface area, filtration volume

A human has about 40×10^9 capillaries with a total exchange surface area of ca. 1000 m^2. Blood flow through the individual organs varies widely at rest and during exercise (see p. 176). Under resting conditions blood flows through only 25–35% of the available capillaries. Some of the venules also participate in fluid exchange (included in above estimate). The amount filtered daily (20 liters) is small in comparison to diffusion (see figure on p. 185).

Transport of substances by diffusion and filtration in the microcirculation and the lymph system, compared to cardiac output (modified after Landis and Pappenheimer, 1963)

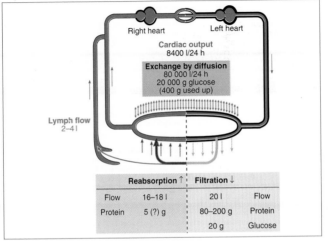

	Reabsorption ↑	Filtration ↓	
Flow	16–18 l	20 l	Flow
Protein	5 (?) g	80–200 g	Protein
		20 g	Glucose

The lymph and its significance

Lymph capillaries (lymphatics) are about as numerous as blood capillaries (exceptions: epidermis, CNS, bones). They begin in the tissues, join to form larger lymph vessels, and after filtration in at least one lymph node open into the venous system, mainly via the thoracic and right lymphatic ducts. The walls of lymphatics are permeable to plasma proteins, fat, sugar and electrolytes. The flow velocity is slow, because under normal conditions the rate of lymph production is only ca. 2 l/24 h; this is the part of the fluid filtered that is not reabsorbed by blood capillaries, and hence consists of interstitial fluid. The average protein content is ca. 20 g/l, with large regional differences (e.g. liver 60 g/l, heart 30 g/l, musculature 20 g/l). In the gastrointestinal tract lipids and other absorbed substances are removed by lymph and delivered to the systemic circulation. In addition to transport, lymph has a very important drainage function. This drainage keeps the tissue 'dry' even when the rate of filtration from blood capillaries into tissue exceeds the rate of reabsorption. But if lymph flow is prevented (inflammation, post-surgery scars), regional lymphatic edema (lymphedema) can result (e.g. in the arm region after removal of the axillary lymph nodes during surgery for breast cancer).

Regulation of regional (local) blood flow

Local regulation of vascular tone

The entire active tension of the smooth muscle of a vessel is called vascular tone. Due to the automatism of smooth muscle fibers (see also p. 87), the arterioles and some large arteries retain basic tone (i.e. remain partly contracted) even after denervation. This applies particularly to the arterioles of organs with highly variable requirements for blood flow (skeletal muscle, exocrine glands), because vasodilation is caused solely by a decrease in tone. Vascular tone is regulated by intrinsic (local) and extrinsic factors (autonomic innervation, hormones). The local factors include:

Factor	Description/action/comments
Temperature	Important in the skin. High ambient temperatures cause vasodilation of the arterioles and veins, visible as reddening of the skin. Cold surroundings cause vasoconstriction, extreme cold produces paradoxical vasodilation
Transmural pressure	High external pressure compresses blood vessels, increasing their resistance (e.g. during muscle contraction or pressure on the skin). Increased intravascular pressure expands blood vessels according to their volume coefficients of elasticity (see p. 178), but in some vessels causes reactive elevation of tone (myogenic autoregulation, Bayliss effect, p. 178)
Local metabolites	Many factors are thought to cause metabolic vasodilation. Among them are hypoxia, acidosis (metabolic or respiratory), ATP, ADP, AMP, adenosine and potassium ions. The importance of these factors varies from organ to organ. When blood flow is interrupted, metabolites accumulate locally, so that when the blood begins to flow again, the flow may overshoot: reactive hyperemia
Local hormones (autacoids)	These are vasoactive substances produced and released locally, and exerting a local action. Among them are histamine (dilates arterioles, constricts veins), bradykinin (dilates arterioles), serotonin, 5-HT (constricts arteries) and the eicosanoids (prostaglandins, thromboxanes and leukotrienes). The F series of prostaglandins (PGF) has a predominantly vasoconstrictor action, the E series (PGE) and prostacyclin (PGI_2) are vasodilators. Thromboxane A_2 is a strong vasoconstrictor, as are the leukotrienes
EDRF	Endothelium-derived relaxing factor, also a local hormone that is released from vascular endothelium and causes vasodilation. Synthesis and release are induced by acetylcholine, also bradykinin, ADP and substance P. Diffuses from the endothelium directly into the underlying muscle cells. It has been demonstrated that EDRF is nitric oxide (NO). Its action is brief (biological half-time 10–40s). Nitroglycerin and other vasodilators also act by releasing NO
Endothelin	Potent, long-lasting (2–3h) vasoconstrictor agent, also a local hormone from the endothelium. The physiological stimulus for its release is not yet known

Neural regulation of vascular tone

Unlike local regulation (preceding page), this has primarily a supraregional role and serves for overall regulation, e.g. of blood pressure or body temperature.

Fiber type	Action/comments
Sympathetic vasoconstrictor	Transmitter is norepinephrine (NE); postsynaptic: α-receptors. Common co-transmitters: ATP, neuropeptide Y. Most important type of vascular innervation, especially for the small arteries and large arterioles. Densely innervated: skin, kidneys, splanchnic region. Sparsely innervated: skeletal muscle, brain. Spontaneous discharge rate of 1–3 imp./s produces the resting tone of the vessels, which exceeds the basic tone (preceding page). Maximal vasoconstriction at ca. 8–10 impulses/s
Sympathetic vasodilator	Transmitter is acetylcholine (ACh); postsynaptic: muscarinic receptors (blocked by atropine); so far known only in skeletal muscles of carnivores (dog, cat). Activation inhibits basic tone. Is activated almost exclusively in stress situations (defense reaction)
Parasympathetic vasodilator	Transmitter is ACh; muscarinic receptors. Common co-transmitter: vasoactive intestinal peptide (VIP). Innervation present in some organs: salivary glands, exocrine pancreas, mucosa of stomach and large intestine, genital erectile tissue, cerebral and coronary arteries/arterioles. No spontaneous activity (resting tone)
Sensory efferent	Release of substance P from activated C afferents has a vasodilator action (axon reflex); unclear physiological role

Hormonal regulation of vascular tone

Several hormones act on the heart and vessels, more in short-term than in long-term adaptation. Their roles are enhanced when neural control is eliminated (e.g. heart transplant) or in pathophysiological situations (e.g. severe bleeding).

Hormone	Action/comments
Epinephrine (E)	Mainly secreted by adrenal medulla (which releases 80% E, 20% NE). Effect similar to that of vasoconstrictor fibers, but weaker. However: in heart and skeletal muscle and in the liver E has a vasodilator action at physiological concentration, because these organs possess (along with α-receptors) many high-affinity β-receptors, activation of which lowers vascular tone. In pharmacological concentrations, the vasoconstricting α action of E predominates
Vasopressin (ADH, see p. 138)	Syn.: antidiuretic hormone. Role in regulation of water balance and hence in long-term adjustment of blood volume to vascular capacity (p. 189). At high concentrations has strongly vasoconstrictor action in most vascular regions; but in brain and heart causes vasodilation by the local release of EDRF (blood redistribution, e.g. during severe bleeding)
Angiotensin II	Extremely strong vasoconstrictor effect. For role in medium-term regulation see p. 189
Atrial natriuretic factor (ANF)	(Also called atriopeptin). Released from atrial musculature of heart; decreases renal vascular resistance and increases renal excretion of salt and water; thereby participates in long-term regulation

Regulation of circulation 1: short-term mechanisms

Rapid regulation of blood pressure (within seconds) involves circulatory reflexes that act as control (negative feedback) circuits. The following sensors participate in circulatory reflexes

Sensor (afferent)	Type of action
Baroreceptors in aortic arch and carotid sinus	Baroreceptor reflex: according to block diagram below. Baroreceptors provide information about heart rate and blood pressure level and amplitude. When stretched they inhibit sympathetic elements and activate the vagus via circulatory centers of the medulla oblongata
Volume (stretch) sensors in atria, and large veins (p. 161)	They detect changes in central venous pressure; correction by vasodilation or vasoconstriction in skeletal musculature (e.g. in orthostatic regulation, blood loss). The same sensors are also involved in long-term volume adjustments. Bainbridge reflex: tachycardia after extreme experimental increase of atrial pressure by intravascular fluid administration
Stretch sensors in ventricles (p. 161)	Have reflex negative inotropic action by way of vagus. Bezold–Jarisch reflex: extreme bradycardia, vasodilation and apnea after chemical stimulation of these sensors by i.v. injection of, e.g. veratrine, nicotine (coronary chemoreflex)
Chemosensors in aortic and carotid bodies	Activated by hypoxia, hypercapnia, acidosis; primarily involved in regulating respiration; maximal stimulation causes peripheral vasoconstriction (except in skin)
Chemosensors in medulla oblongata	In extreme central hypoxia, hypercapnia and acidosis, excitation of these sensors elicits an ischemia reaction (Cushing reaction): extreme sympathetic excitation with vasoconstriction and great rise in blood pressure (interpreted as an emergency reaction to improve supply of blood to the brain)

Baroreceptor reflex: example of operation of the short-term control systems

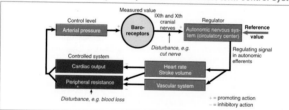

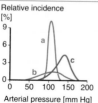

Relative incidence [%]

Arterial pressure [mm Hg]

Frequency distribution of mean blood pressure in the dog under normal conditions (curve a) and a few days after denervation of the arterial baroreceptors (b) and also of the cardiopulmonary stretch or volume sensors (c). In the first case only the blood pressure fluctuations increase, but in the last the mean pressure also rises ('release' hypertension) (from data of Cowley *et al.*, 1989)

Regulation of circulation 2: medium-term mechanisms

Onset of action within minutes, fully developed after hours

Mechanism	Description/comments
Transcapillary volume shifts	Changes in blood volume by net transcapillary fluid flow. This is brought about by changes in capillary effective filtration pressure (for mechanisms see p. 184)
Stress relaxation of vessels	Slow change in distensibility (delayed compliance) of venous vessels after rapid filling or emptying, so that despite a large change in volume the intravascular pressure returns towards initial levels
Renin–angiotensin system (RAS)	Renin is released in response to reductions in renal perfusion pressure. Renin converts angiotensinogen to angiotensin I, which is converted to angiotensin II by the angiotensin-converting enzyme. Angiotensin II has a strong vasoconstrictor action, excites the sympathetic system, stimulates secretion of aldosterone (see below) and elicits thirst

Regulation of circulation 3: long-term mechanisms

These mechanisms control the extracellular volume and hence the filling of the vascular system (i.e. primarily the central venous pressure, see p. 182)

Mechanism	Description/comments	
Renal volume regulation system	Elevated arterial pressure causes increased loss of salt and water in the urine and thereby lowers central venous pressure (reduces venous supply), which in turn reduces cardiac output and lowers blood pressure. Conversely, a drop in blood pressure reduces salt and water excretion, etc. Small changes in arterial pressure are associated with considerable changes in renal salt and water excretion	Relative fluid excretion or fluid uptake [x normal] Excretion Excretion (isolated kidney) Normal uptake Arterial pressure [mm Hg] (after Guyton, 1976)
Vasopressin system	For mechanism of action of vasopressin (antidiuretic hormone, ADH) see p. 138. ADH controls reabsorption of water in the collecting duct of the nephron (p. 256) and its secretion is stimulated by elevated plasma osmolality. ADH secretion is also increased by large falls in extracellular fluid volume	
Aldosterone system	Aldosterone increases total body salt and water content by increasing tubule reabsorption of Na^+ and water (with simultaneous secretion of H^+ and K^+, p. 255); in this respect is also part of the above volume regulation system. Strongest stimulus of aldosterone secretion is angiotensin II (see above)	

Special features of subdivisions of the circulatory system

The differences in blood flow and oxygen consumption of individual organs can be found on p. 176. Here some other important aspects of organ circulation are summarized (organs in alphabetical order)

Subdivision	Special features/comments
Brain	Marked local metabolic regulation of blood flow while total cerebral blood flow is kept nearly constant (high sensitivity to P_{CO_2}). See also p. 115. Blood–brain barrier function of capillaries is another special feature
Kidney	Marked autoregulation keeps blood flow practically constant over wide range of pressures; large blood flow in renal cortex, little in medulla. Two capillary beds in series: afferent arterioles → glomerular capillary bed (ca. 60 mmHg) → efferent arterioles → peritubular capillary bed (ca. 13 mmHg)
Liver, portal vein	Blood from intestine, pancreas and spleen (superior mesenteric and splenic arteries) flows through the portal veins to the liver, which also receives blood from the hepatic artery (p. 174); common outflow in hepatic veins. Sympathetic splanchnic nerves supply mesenteric, pancreatic, splenic and hepatic vessels, together called the splanchnic region (p. 176). Marked local metabolic regulation, which can suppress vasoconstrictor tone (autoregulatory escape). Liver also serves as blood reservoir: vasoconstriction can temporarily release 50% of the hepatic blood volume (700 ml)
Lung	(Pressures and volumes on pp. 175, 176.) Pulmonary vessels are subject to strong hydrostatic influences because of the low arterial pressures; in upright position, blood flow through apical regions is much less than at the basal regions. Central blood volume consists of blood in lungs (440 ml, p. 175) plus diastolic volume of left ventricle (together 600–650 ml), of which 50% can be mobilized immediately

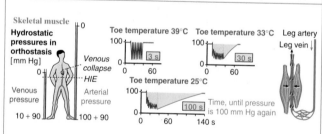

Blood flow depends on muscle work; in addition to local metabolic and nervous vasodilation, the nature of the contraction also affects blood flow: postural work (isometric contraction) reduces, and rhythmic contraction assists blood flow by a muscle pump effect. Figure shows interactive effects of hydrostatic pressure, muscle pump and temperature-dependent arteriole diameter on blood flow through leg muscles during walking. (after Henry and Gauer, 1950)

| Skin | Main function is to contribute to thermoregulation (by variation of cutaneous blood flow, also by sweating: cholinergic sweat gland innervation); under thermoneutral conditions there is high vasoconstrictor tone. Thermally induced increase in blood flow opens many arteriovenous anastomoses. Subpapillary venous plexus also serves as blood reservoir (capacity 1500 ml) |

Fetal circulation

Diagram of the fetal circulation (left) and the circulation after birth (right). The O_2-containing blood is shown in light red; the graded darkening corresponds to decreasing O_2 content

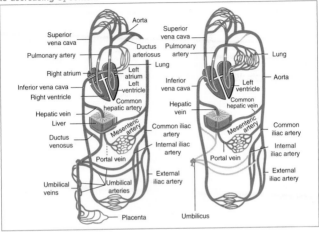

Special features of the fetal circulation

Ductus venosus	Carries most of the oxygenated blood from the umbilical vein into the inferior vena cava. The rest enters the latter by way of the liver, the best-supplied fetal organ, with the highest metabolic rate
Foramen ovale	Through this opening between the atria most of the venous mixed blood from the inferior vena cava passes into the left atrium. The blood from the superior vena cava predominantly enters the right ventricle. In the right atrium of the fetus there is thus a cross-current of blood
Ductus arteriosus	Connection between pulmonary artery and aorta, through which most of the right cardiac output enters the aorta. The rest supplies the still-collapsed lung (higher pressure in right than in left ventricle)
Umbilical arteries	Arise from the internal iliac arteries on both sides. That is, there are two umbilical arteries but only one umbilical vein
Changes at birth	Expansion of the lungs increases pulmonary blood flow and hence the venous return to the left atrium. Pressure in the right atrium falls simultaneously due to absence of placental blood, so that the foramen ovale closes like a valve. Ductus arteriosus is slowly obliterated

Synopsis of important functional data on respiration

Average values for a healthy young man (body surface area 1.7 m²) at physical rest (data from Thews, 1989)

Lung volumes	
Total capacity	6 liters
Vital capacity	4.5 liters
Funct. residual capacity	2.4 liters
Tidal volume	0.5 liters
Dead-space volume	0.15 liters

Ventilation	
Respiratory rate	14 min⁻¹
Minute volume	7 l/min
Alveolar ventilation	5 l/min
Dead-space ventilation	2 l/min

Gas exchange	
O_2 uptake	280 ml/min
CO_2 release	230 ml/min
Respiratory quotient	0.82
O_2 diffusion cap.	30 ml·min⁻¹·mmHg⁻¹ (230 ml·min⁻¹·kPa⁻¹)
Contact time	0.3 s

Respiratory mechanics	
Intrapleural pressures at:	
end of expiration	−5 cmH₂O (−0.5 kPa)
end of inspiration	−8 cmH₂O (−0.8 kPa)
Compliance of the lung	0.2 l/cmH₂O (2 l/kPa)
Compliance of the thorax	0.2 l/cmH₂O (2 l/kPa)
Compliance of lung plus thorax	0.1 l/cmH₂O (1 l/kPa)
Resistance	2 cmH₂O·s·l⁻¹ (0.2 kPa·s·l⁻¹)

Function tests	
Relative forced expiratory volume	75%
Max. volume flow	10 l/s
Maximum breathing capacity	150 l/min

Perfusion relationships	
Alveolar ventilation/ perfusion	0.9
Shunt perfusion/ total perfusion	0.02

Spirometric measurement of lung volumes and capacities (sets of volumes). Residual volume and (for normal expiration) functional residual capacity must be found by different means (see p. 195)

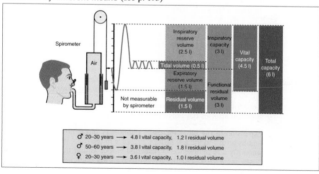

♂ 20–30 years → 4.8 l vital capacity, 1.2 l residual volume
♂ 50–60 years → 3.8 l vital capacity, 1.8 l residual volume
♀ 20–30 years → 3.6 l vital capacity, 1.0 l residual volume

Physical fundamentals

Conditions for measurement of gas volumes

Name	Definition/comments	T(K)	p (mmHg)
STPD	**S**tandard **t**emperature, **p**ressure, **d**ry; 0°C (273 K), pressure 760 mmHg (101 kPa), dry (water vapor pressure zero). These are the physical standard conditions	273	760
BTPS	**B**ody **t**emperature, **p**ressure, **s**aturated: 37°C, ambient pressure (barometric pressure, p_B), complete saturation with water vapor (i.e. 47 mmHg at 37°C, see below)	310	$p_B - 47$
ATPS	**A**mbient **t**emperature, **p**ressure, **s**aturated: ambient temperature T_a, ambient air pressure p_a, complete saturation with water vapor p_{H_2O} (value depends on T_a). These are the 'spirometer conditions'	T_a	$p_B - p_{H_2O}$

Behavior of ideal gases when ambient conditions are changed; p, pressure; V, volume; n, number of moles; R, gas constant; T, absolute temperature (t°C + 273)

Gas law	Definition/description/comments
pV = nRT	**Ideal gas equation**, can be applied to pure gases (e.g. O_2) and gas mixtures (e.g. air)
pV = constant	**Boyle–Mariotte law** (for constant T and n). Special case of the ideal gas equation
V/T = constant	**Gay-Lussac law** (for constant p and n). Special case of the ideal gas equation. When T = 0 (−273°C), V = 0. From there increases linearly by 1/273 per °C
$p = p_1 + p_2 \ldots + p_n$	**Dalton law**: law of additivity of partial pressures p_1 to p_n. Valid for gas mixtures that contain non-ideal gases, such as alveolar air

Practically all physiologically important gases, including carbon dioxide, obey the ideal gas law to a good approximation. There is one important exception: water vapor, see below.

The partial pressure of water vapor has as its upper limit its temperature-dependent saturation value (*unlike* ideal gases). Examples of saturation pressure:

Temperature (°C)	0	15	20	25	30	34	37	40	100
H_2O pressure (mmHg)	5	13	18	24	32	40	47	55	760

Air in the alveoli is saturated with water vapor. As a consequence, the H_2O partial pressure there is 47 mmHg (BTPS conditions). At room temperature (ATPS conditions) it is considerably lower, which according to the gas laws causes the volume to be reduced. Withdrawal of water and lowering the temperature to 0°C (STPD conditions) further reduces the volume. The ratio of V_{STPD} to V_{ATPS} under normal ATPS conditions is 0.89, and that of V_{STPD} to V_{BTPS} is 0.81.

Ventilation of the lungs

Respiratory frequency f and tidal volume V_E depend on age and size; normal values at rest are (from Müller, 1991)

Rule of thumb for V_E	Age group	f (per min)	V_E (ml)
V_E (ml) =	Newborns	40–50	20–35
in children:	Infants	30–40	40–100
weight (kg) × 10	Small children	20–30	150–290
	School children	16–20	300–400
in adults:	Adolescents	14–16	300–500
weight (kg) × (10–15)	Adults	10–14	500–1000

Expiratory minute volume $\dot{V}_E$, alveolar ventilation $\dot{V}_A$ and dead-space ventilation $\dot{V}_D$

The minute volume is the product of tidal volume and respiratory frequency (values above and in Table p. 192). For adults at rest $\dot{V}_E$ and $V_E f = 0.5 \times 14 = 7 l/min$. Under strenuous exercise, this value can rise to 120 l/min. The part of $\dot{V}_E$ that actually ventilates the alveoli is called the alveolar ventilation; the rest is the dead-space ventilation: hence $\dot{V}_E = \dot{V}_A + \dot{V}_D$. The dead-space V_D has a volume of about 150 ml, or 30% of the resting V_E (see p. 195)

Lung volumes and capacities: definitions and comments (for spirometric measurement and normal values see p. 192)

1. Tidal volume: Normal inspiratory or expiratory volume; values given above. Normally the expired volume is somewhat smaller than that inspired, because more O_2 is taken up than CO_2 is given off (respiratory quotient RQ < 1). By convention the expiratory volumes are given, hence the subscript E.

2. Inspiratory reserve volume: Volume that can still be inhaled after a normal inspiration; ca. 2.5 liters.

3. Expiratory reserve volume: Volume that can still be exhaled after a normal expiration; ca. 1.5 liters. (Hence the 'medium position' for normal respiration is in fact below the 'middle'.)

4. Residual volume: Volume remaining in the lungs after maximal expiration; ca. 1.5 liters. Increases with age, because (a) the lungs become less elastic and (b) the thorax stiffens, so that the vital capacity (5) decreases.

5. Vital capacity (VC): Maximal volume that can be exhaled after maximal inspiration, i.e. the sum of 1, 2 and 3. For guideline values (age- and sex-dependent) see figure. The VC is no fully exploited even during maximal breathing. Endurance training increases VC.

6. Inspiratory capacity: Maximal volume that can be inhaled after normal expiration, i.e. the sum of 1 and 2.

7. Functional residual capacity, FRC: Volume remaining in the lungs after normal expiration, i.e. the sum of 3 and 4. The FRC 'buffers' the fluctuations of the O_2 and CO_2 partial pressures during the respiratory cycle, because its volume of ca. 3 liters is several times greater than the tidal volume.

8. Total capacity: Volume within the lungs after maximal inspiration, i.e. the sum of 4 and 5.

Anatomical and functional dead space V_D ($V_E = V_A + V_D$; see p. 193)

Definitions:	Calculation:
Anatomical V_D: The airway components in which no gas exchange occurs (nose, mouth, pharynx, larynx, trachea, bronchi, bronchioles); ca. 150 ml Physiological (functional) V_D: Anatomical V_D plus alveoli that are ventilated but not perfused; the latter component is neglibible in healthy people	Employs the Bohr formula after measurement of the CO_2 fractions in the expired air ($F_{E_{CO_2}}$) and in the alveolar air ($F_{A_{CO_2}}$): $$\frac{V_D}{V_E} = \frac{F_{A_{CO_2}} - F_{E_{CO_2}}}{F_{A_{CO_2}}}$$

Measurement of functional residual capacity (FRC) and residual volume

FRC and residual volume can only be measured indirectly. Usually what is measured is the dilution of relatively insoluble inert test gases, often the foreign gas helium (helium dilution method) or the nitrogen already present in the lungs (nitrogen washout method); the nitrogen is 'washed out' by breathing pure oxygen. A third method is to use the body plethysmograph. With the subject in a rigid chamber, the pressure changes in chamber and mouth are measured as the subject tries to breathe in and out with airways closed; the lung volume is then calculated by the Boyle–Mariotte law ($p \cdot V$ = constant; see p. 194).

Forms of breathing (ventilation) 1: in terms of alveolar P_{CO_2}

Ventilation form	Description/comments
Normoventilation	Normal breathing, in which the CO_2 partial pressure in the alveoli (and capillaries) is kept near 40 mmHg (normal range 35–45 mmHg)
Hyperventilation	Ventilation increased beyond metabolic requirements. Causes a reduction of the alveolar and arterial P_{CO_2} (hypocapnia)
Hypoventilation	Ventilation reduced below metabolic requirements. Causes an increase in the alveolar and arterial P_{CO_2} (hypercapnia)
Increased ventilation	Ventilation above the resting level but with normal P_{CO_2}, e.g. during exercise. Syn.: polypnea, hyperpnea, see below

Forms of breathing (ventilation) 2: in terms of minute volume, respiratory frequency and clinical observations

Ventilation form	Description/comments
Eupnea	Normal resting respiration
Hyperpnea or polypnea	Deeper breathing (increased minute volume) with or without increase in respiratory frequency (increased ventilation, see above)
Hypopnea	Reduced minute volume
Tachypnea	Increase in respiratory frequency above the normal frequency of 14–18/min
Bradypnea	Decrease in respiratory frequency
Apnea	Temporary cessation of breathing, mainly caused by decreased arterial P_{CO_2} (e.g. after voluntary hyperventilation)
Dyspnea	Difficult breathing with subjective feeling of shortness of breath
Orthopnea	Severe dyspnea in left heart failure, forces patient to sit up
Asphyxia	Cessation or reduction of breathing when respiratory centers are damaged, leads to hypoxia and hypercapnia

Respiratory mechanics

Lung elasticity, intrapleural (intrathoracic) pressure, pneumothorax

Two forces pull the lung in the direction of the hilum: (1) the elasticity of the lung tissue and (2) the surface tension of the liquid layer in the alveoli (the latter is reduced by the presence of surfactants). Because of these tensile forces, the pressure in the pleural space is lower than the ambient air pressure; this intrapleural (syn.: intrathoracic) pressure is largest at the end of inspiration, because at this time the lung is more distended (for values p. 192). It can be measured with sufficient accuracy as the esophageal pressure (with a probe in the esophagus). If the space between the pleural layers is connected to the atmosphere (probe, injury), the lung contracts toward the hilum and air enters the pleural space, producing pneumothorax. The lung on the affected side no longer follows the breathing movements.

Elastic resistances in the lungs

During inspiration the pressure in the airways and the alveoli must fall below the ambient air pressure (or be lower than an above-atmospheric pressure generated with a respiratory pump), in order for air to flow into the lungs. Conversely, the intrapulmonary pressure must be higher than the ambient air pressure for air to flow out of the lungs. The relation between intrapulmonary pressure p_{pul} and the change in lung volume ΔV for the whole respiratory system can be found with a respiratory pump and expressed as a static volume–pressure curve. The slope of this curve is called the compliance C_{L+T} of the respiratory system, the subscript L + T indicating that it is the combined compliance of lung and thorax. For normal breathing movements it is 0.1 l/cmH$_2$O (see also Table p. 192). Therefore to inhale the tidal volume of 0.5 liter, a pressure difference of ca. 5 cmH$_2$O must be generated.

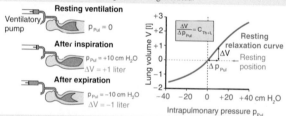

The static volume–pressure curves of lung and thorax can also be assessed separately, so each compliance can be calculated. The two values are about the same (see Table p. 192). The relationships between C_{L+T}, the compliance of the lung C_L and that of the thorax C_T are as follows (where P_{pul} is the intrapulmonary pressure and P_{pleu} is the intrapleural pressure):

$$C_{L+T} = \frac{\Delta V}{\Delta p_{pul}} \qquad C_L = \frac{\Delta V}{\Delta \left(p_{pul} - p_{pleu}\right)} \qquad C_T = \frac{\Delta V}{\Delta p_{pleu}}$$

The three equations give the following relationship:

$$\frac{1}{C_{L+T}} = \frac{1}{C_L} + \frac{1}{C_T}$$

Because the compliance C is a measure of elasticity, 1/C is a measure or the stiffness or elastic resistance of the entire respiratory system, an additive combination of the resistances of thorax and lung.

Viscous (non-elastic) resistances

90% of the viscous resistance is the airway resistance R to the flow of air into and out of the lung, and 10% is tissue resistance (tissue friction and non-elastic deformation of the tissue in thoracic and abdominal cavities). To a first approximation, the volume flow $\dot{V}$ is proportional to the driving pressure difference Δp (Hagen–Poiseuille law), i.e.

$$\dot{V} = \frac{\Delta p}{R} = \frac{P_{pul}}{R}$$

where the resistance to flow R depends on the cross-sectional area and length of the tube and on viscosity. Rearrangement gives

$$R = \frac{\Delta p}{\dot{V}} = \frac{P_{pul}}{\dot{V}}$$

(value on p. 192, measured by body plethysmography)

Pressure–volume relations during the respiratory cycle

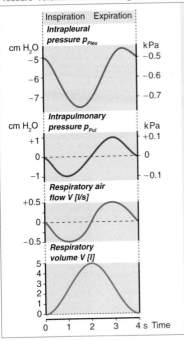

At the transition from the preceding expiration to the next inspiration, the thorax is briefly at rest. The intrapleural pressure is negative (see above), the intrapulmonary pressure is zero. During inspiration the thorax is expanded by the respiratory muscles. As a result, the intrapleural pressure becomes more negative (values on p. 192). The expansion of the alveoli causes the intrapulmonary pressure to fall as well. The fall in intrapulmonary pressure causes air to flow into the lung down the pressure gradient. The pressure difference falls progressively during inspiration, becoming eventually zero. At this point inspiration is completed.

During expiration the elastic forces in the thorax and lung (under some conditions together with the expiratory muscles) pull the thorax back to its initial position. At first the intrapulmonary pressure rises, i.e. a pressure gradient builds up from inside to outside. Air now escapes from the lungs until the pressure gradient is abolished. In parallel, the intrapleural pressure becomes less negative. At the end of expiration, an entire respiratory cycle has been completed.

Pressure–volume diagram and respiratory work

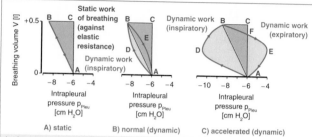

A) static B) normal (dynamic) C) accelerated (dynamic)

A plot of respiratory volumes as a function of intrapleural (intrathoracic) pressures is called the pressure–volume diagram of the lung (comparable to the Frank–Starling diagram for the heart, p. 169). The respiratory work is given (in complete analogy to the work of the heart, p. 172) by the product of pressure and tidal volume. When plotted in rectangular coordinates, work is represented by an area (**A**: ABCA, **B**: ADBCA; **C**: ADBCFEA). Under (purely theoretical) static conditions work must be done only against the elastic forces during inspiration (expiration occurring by passive elastic restoration). Even for normal breathing no active work need be performed for expiration. At rest 2% of the O_2 uptake is needed for the contractile work of the respiratory muscles; during heavy exercise this value can rise to 20%.

Dynamic tests of respiratory function (normal values in the Table on p. 192)

Test	Definition/description/comments
Tiffeneau test	Measurement of the volume that can be expired in 1 s after maximal inspiration (forced expiratory volume). Usually expressed as a percentage of vital capacity.
Maximal volume flow	Measurement (with a pneumotachograph) of the maximal volume flow achievable during expiration starting from maximal inspiration
Maximum breathing capacity	Measurement of the tidal volume during voluntary maximal forced hyperventilation. Test should be carried out for only ca. 10 s, to prevent respiratory alkalosis, but result is expressed per min

Restrictive and obstructive ventilatory disorders and their diagnosis

Disorder	Definition/description/comments
Restrictive	All disorders in which the lung's ability to expand is impaired, e.g. pulmonary fibrosis, pleural thickening. Compliance and vital capacity are reduced. Maximum breathing capacity decreases (but it decreases also in obstructive disorders, hence it is useless for differential diagnosis)
Obstructive	Narrowing of the conducting airways, i.e. increase in resistance to flow, e.g. by mucus in the bronchioles, bronchial asthma (spasms of the bronchial musculature). Diagnosis with Tiffeneau test: result below normal in obstructive disorders due to increased resistance. The maximal expiratory volume flow is also far below normal

Gas exchange in the lungs

Inspiratory, alveolar and expiratory fractions and partial pressures of the respiratory gases during resting respiration at sea level

	Fractions		Partial pressures	
	O_2	CO_2	O_2	CO_2
Inspired air	0.209 (20.9 vol. %)	0.003 (0.3 vol. %)	150 mmHg (20 kPa)	0.2 mmHg (0.03 kPa)
Alveolar gas mixture	0.14 (14 vol. %)	0.056 (5.6 vol. %)	100 mmHg (13.3 kPa)	40 mmHg (5.3 kPa)
Expired mixture	0.16 (16 vol. %)	0.04 (0.4 vol. %)	114 mmHg (15.2 kPa)	29 mmHg (3.9 kPa)

Measurement of the alveolar and expired mixtures by the volumetric gas analysis method of Scholander (absorption of CO_2 by KOH, reduction of O_2 to H_2O by sodium dithionite) or continuous measurement with modern analytical methods (e.g. CO_2 by infrared absorption measurement). For all partial pressure calculations the water vapor pressure must be subtracted (47 mmHg in alveolar gas mixture, see p. 194) because it varies greatly in the outside air. Hyperventilation causes the alveolar Po_2 to rise and the Pco_2 to fall (the latter is important: danger of respiratory alkalosis). Hypoventilation has the opposite effects (danger of hypoxia, shortness of breath due to increased Pco_2).

Temporal variation of respiratory gas partial pressures in the alveolar air during a respiratory cycle (modified after Piiper, 1975)

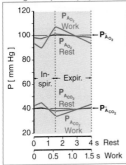

The respiratory gas partial pressures P_A vary only slightly about the temporal mean values $\bar{P}_A$ because the ventilation coefficient (ratio of fresh alveolar air [at rest] 350 ml] to the air remaining in the lung, i.e. the functional residual capacity [at rest 2.4 liters, see p. 192]) is around 14% (sufficiently close approximate value: 10%). During exercise, the values fluctuate somewhat more, in a more rapid rhythm. At the beginning of inspiration 'old' alveolar air is first reinspired from the dead space, so that P_{Ao_2} at first falls further (P_{Aco_2} rises), until fresh air reaches the alveoli. The changes during expiration are brought about by the continuous diffusion of O_2 from alveoli to blood and by the diffusion of CO_2 in the opposite direction.

Respiratory parameters, O_2 uptake and CO_2 release at rest (values also on p. 192) and during heavy exercise

	Parameter	Rest	Heavy exercise
a	O_2 uptake	280 ml/min	3000 ml/min
b	CO_2 release	230 ml/min	2850 ml/min
c	Respiratory quotient (b/a)	0.82	0.95
d	Respiratory frequency	14/min	40/min
e	Tidal volume	0.5 liter	2 liters
f	Respiratory minute volume (d · e)	7 l/min	80 l/min

Alveolar diffusion of respiratory gases, O₂ diffusion capacity

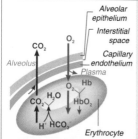

Alveolar epithelium

Interstitial space

Capillary endothelium

CO_2 O_2

Alveolus

Plasma

O_2
Hb

CO_2 H_2O

HbO_2

H^+ HCO_3^-

Erythrocyte

O_2 and CO_2 are exchanged across a large area in the lung (total alveolar surface area 80 m²) and a short diffusion path (ca. 1 μm, see Fig.), namely 2 thin layers, capillary endothelium and alveolar epithelium. Venous blood has $P_{O_2} = 40$ mmHg, and $P_{CO_2} = 46$ mmHg. The alveolar pressures are 100 and 40 mmHg, respectively. The partial pressure differences (60 mmHg for O_2, 6 mmHg for CO_2) represent the driving forces for pulmonary gas exchange. Because the diffusion conductivity (Krogh's diffusion coefficient) for CO_2 is 23 times higher than that for O_2, the small partial pressure difference (6 mmHg) suffices for adequate CO_2 diffusion.

From the arterial to the venous end of the alveolar capillary, the partial pressure differences decrease progressively, and at the end of the capillaries they are zero, i.e. the arterial blood has the same partial pressures as the gas mixture in the corresponding alveoli. The mean O_2 partial pressure difference (alveoli–arterial blood) is 10 mmHg (see below). From this value one can calculate, for an O_2 uptake of ca. 300 ml/min, an O_2 diffusion capacity of 30 ml/min per mmHg (see Table p. 192).

Pulmonary perfusion (blood flow) and contact time; ventilation–perfusion ratio; shunt perfusion (for normal values see Table p. 192)

The time an erythrocyte takes to pass through a pulmonary capillary, i.e. its contact time with the alveolar gases, is 0.3 s. The total volume of lung capillaries at rest is about 100 ml; 50% of the capillaries in the lung are perfused at rest. During heavy exercise the rest of the capillaries open and all of them enlarge their diameter; accordingly, both capillary volume and alveolar exchange area increase. The ratio of alveolar ventilation. $\dot{V}_A$ (5 l/min) to pulmonary perfusion $\dot{Q}$ (blood flow through lung, ca. 5 l/min) is about $\dot{V}_A/\dot{Q} = 0.9$ at rest (for regional differences, see below). Only 2% of the blood in the lung flows through arteriovenous short circuits (shunt perfusion).

Effects of regional inhomogeneities in the lung on arterialization of the blood
(fig. after Thews, 1989)

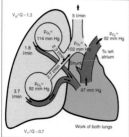

$V_A/Q \sim 1.3$

5 l/min

$P_{O_2} =$ 114 mm Hg

$P_{O_2} =$ 102 mm Hg

$P_{O_2} =$ 92 mm Hg

1.8 l/min

To left atrium

Shunt

$P_{O_2} =$ 92 mm Hg

97 mm Hg

3.7 l/min

Work of both lungs

$V_A/Q \sim 0.7$

In the upright position, the base of the lung is more perfused than the apex, due to the hydrostatic pressure difference (see p. 183). Hence in the base, $V_A/Q < 1$, so that the arterial blood is less saturated with O_2 than that in the apex of the lung (see Fig.) Blood from the two regions mixes to give a P_{O_2} of 97 mmHg, which is reduced to 92 mmHg because of shunted blood. The final arterial P_{O_2} is thus ca. 10 mmHg below the alveolar P_{O_2} (further decrease in old age, only 70 mmHg at age 70).

The control of breathing

Survey of specific and nonspecific factors that drive respiration
(modified from Thews, 1989)

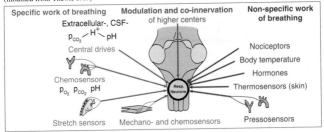

Specific work of breathing — Modulation and co-innervation of higher centers — Non-specific work of breathing

Extracellular-, CSF-
p_{CO_2} — H^+ — pH
Central drives

Chemosensors
p_{O_2} p_{CO_2} pH

Stretch sensors — Mechano- and chemosensors

Resp. Neurone

Nociceptors
Body temperature
Hormones
Thermosensors (skin)

Pressosensors

Specific driving factors 1: chemical factors, associated chemoreceptors
(Figure modified from G. Thews, based on data of H.P. Koepchen)

Action of CO_2	Normally the dominant stimulus; see CO_2 response curve in **A** of the Fig. Action has upper limit at V_E of 75 l/min
Action of H^+	A decrease below the normal value of pH 7.4 slightly enhances respiration. Red curve in **B** shows effect of acidosis by increase in non-volatile acids (smaller effect because blood Pco_2 falls as V_E increases), black curve shows pH effect when Pco_2 is kept constant experimentally
Action of O_2	Normally relevant only at high altitudes, see pp. 233, 235. Red curve in **C** shows effect of changes in Po_2 (again small because blood Pco_2 falls with hyperventilation), black curve shows 'pure' Po_2 effect

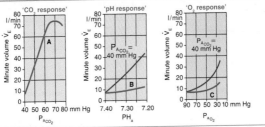

Peripheral chemoreceptors	Locations: Carotid bodies (IXth cranial nerve), aortic bodies (Xth cranial nerve). Respond with increased activity to increase of H^+ or Pco_2, decrease of O_2. O_2 effects are mediated exclusively by the peripheral chemoreceptors
Central chemoreceptors	Locations: Areas on the ventral surface of the brainstem. Chief responsibility for H^+ and CO_2 effects. Crucial factor seems to be the H^+ concentration of the extracellular fluid

Specific driving factors 2: mechanoreceptors in the lungs (Hering–Breuer reflex), the respiratory muscles and the other skeletal muscles

Stretch receptors in the lungs	Mechanoreceptors of the lung parenchyma and the airways are activated by the expansion of the lungs during inspiration and reflexly inhibit further inspiration: Hering–Breuer reflex. Limits the amplitude of respiratory excursions. Afferent fibers run in the vagus nerve (Xth). After vagotomy, breathing becomes deeper and slower
Mechanoreceptors in the respiratory muscles	Muscle spindles and Golgi tendon organs of the respiratory muscles contribute to the regulation of breathing by their mono- and polysynaptic spinal and supraspinal reflex arcs, helping to adjust respiratory mechanics to the prevailing mechanical conditions
Receptors in the skeletal muscles	Mechanoreceptors with thin afferent fibers (ergosensors) and chemosensitive muscle afferent fibers (metabosensors) are excited during intense muscle work. The afferent discharges appear to contribute to the reflex increase in tidal volume

Central co-innervation

At the beginning of physical exercise, the tidal volume has already increased far more than explained by changes in blood chemistry. The best explanation is that the respiratory centers receive from the cortical motor centers an 'efference copy' of the commands to voluntary muscles, so that breathing is adjusted in 'anticipation' of the forthcoming requirements (feed-forward mechanism).

Non-specific respiratory drives

General term for influences that do not primarily function in the control of breathing. Examples in figure, e.g. noxious and thermal stimulation of the skin, activation of baroreceptors in the carotid sinus, release of epinephrine.

Pathological breathing (schematic) (modified from Thews, 1989)

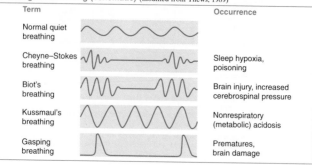

Term		Occurrence
Normal quiet breathing		
Cheyne–Stokes breathing		Sleep hypoxia, poisoning
Biot's breathing		Brain injury, increased cerebrospinal pressure
Kussmaul's breathing		Nonrespiratory (metabolic) acidosis
Gasping breathing		Prematures, brain damage

Localization of the respiratory neurons ('respiratory centers') in the brainstem of the cat, by electrophysiological recording (after Richter, 1986)

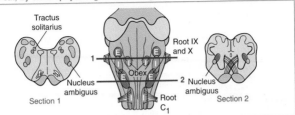

Inspiratory neurons (I): dorsal group next to the nuclear region of the solitary tract, ventral group near the nucleus ambiguus and in cervical region (C_{1-2}).

Expiratory neurons (E): dorsal group next to the nucleus ambiguus, ventral group at the retrofacial nucleus.

Discharge patterns of phasic respiratory neurons (schematic, based on H.P. Koepchen)

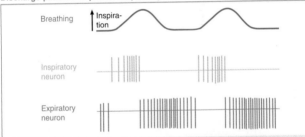

In addition to the activity patterns shown here there are many others, on the basis of which six classes of respiratory neurons are currently distinguished. It seems certain that there are pacemaker neurons for breathing, which (like the pacemaker cells of the heart) are spontaneously rhythmically active and thereby provide the basic pattern of the respiratory rhythm. This rhythm persists after all afferent inputs have been cut off. Normally it is modified by these inputs, according to physiological needs.

Inspiratory neurons seem to be considerably more numerous and diverse than expiratory neurons (reflects the fact that inspiration is a more active process). Reciprocal postsynaptic inhibitory potentials (IPSPs) or connections between the neurons are just as common as or more common than postsynaptic excitatory potentials (EPSPs) or connections.

Properties of the chromoprotein hemoglobin (Hb, molecular weight 64 500)

Structure/function	Description/comments
Globin	Protein comprising 4 polypeptide chains, of which two are α-chains, each with 141 amino acids (AA), and two are β-chains, each with 146 AA. In fetal Hb (HbF) the β-chains are replaced by γ-chains with slightly different AA sequence
Heme	Pigment (4% of whole molecule): protoprophyrin with central Fe^{2+} atom, one heme for each polypeptide chain, i.e. four per Hb molecule
Hb/HbO_2	O_2 is reversibly bound to the heme of Hb with no change in valence; O_2 binding is called oxygenation, and the release of O_2 deoxygenation
Methemoglobin	Compound formed by genuine oxidation of heme, with conversion of iron from bivalent to trivalent. Pathological state, because it makes hemoglobin unavailable for O_2 transport
Colors of Hb and HbO_2	Result from differential light absorption: blue light is absorbed by Hb and red light is transmitted; therefore Hb appears dark red, while HbO_2 absorbs still more blue and looks light red

Hb content and mean corpuscular hemoglobin of human blood
(modified from Thews, 1989)

Hemoglobin concentration in male and female adults and in the newborn	

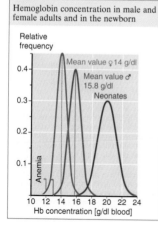

The Hb concentrations were found by extinction measurement (spectral photometry, cyanmethemoglobin method and application of the Lambert–Beer law).

The mean Hb content of a single erythrocyte (MCH, mean corpuscular hemoglobin) in both men and women is 31 pg (normochromic). Blood with lower MCH is hypochromic, that with higher MCH is hyperchromic.

MCH is found by dividing the Hb content by the number of erythrocytes, hence in men

$$MCH = \frac{158\,g}{5.1 \cdot 10^{12}} = 31 \times 10^{-12}\,g = 31\,pg$$

and in women

$$MCH = \frac{140\,g}{4.6 \cdot 10^{12}} = 31 \times 10^{-12}\,g = 31\,pg$$

O₂ transport function of the blood

Characteristic values for O₂ transport and binding capacity of the blood

Term	Definition/description/comments
Solubility	Directly proportional to the gas partial pressure P_{gas}, according to the Henry–Dalton law $C_{gas} = \alpha p_{gas}$. The proportionality factor α is the Bunsen solubility coefficient. At normal body temperature, 1 ml blood contains 0.003 ml O_2
Reaction equation	1 mol Hb can bind at most 4 mol O_2 (because of the tetrameric structure of the Hb molecule), $Hb + 4O_2 \rightleftharpoons Hb(O_2)_4$. Hence 1 g Hb binds maximally 1.39 ml O_2. In practice the maximum is lower, because small amounts of Hb are inactive; see Hüfner's number, below
Hüfner's number	1 g Hb *in vivo* binds 1.34 ml O_2. The product of Hüfner's number and Hb concentration gives the maximal O_2 binding capacity: $[O_2]_{max} = 1.34$ (ml O_2/g Hb) $\cdot 150$ (g Hb/l blood) $= 0.20$ liters O_2 per liter of blood). This relationship applies strictly only to O_2 partial pressure $>300\,mmHg$. It is approximately reached in the lung due to the S shape of the binding curve (see below)

Oxygen dissociation curves of the blood

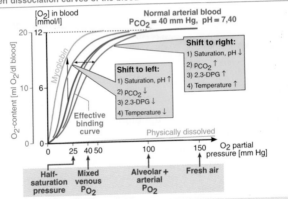

The oxygenated fraction of Hb is called the O_2 saturation, S_{O_2}. S_{O_2} and hence the O_2 content of the blood is directly proportional to O_2 partial pressure. The S shape of the curve for S_{O_2} is due to differences in O_2 affinity of the 4 heme molecules in Hb (intermediate compound hypothesis). The half-saturation pressure P_{50} is a P_{O_2} of 26 mmHg. Biological significance: Flattening of the curve at high partial pressures ensures adequate saturation in the lung, while the steep slope in the middle allows much O_2 to be released to the tissue with slight partial-pressure reductions. Shift to the right means that the O_2 affinity of Hb decreases, so higher O_2 partial pressures are required to obtain a given S_{O_2}. Shift to left means that the O_2 affinity increases, so more O_2 is bound at a given O_2 partial pressure. Extreme example: CO-Hb. Details on next page.

Factors influencing the O_2 dissociation curve; other globins

Factor	Action/comments
pH, CO_2, Bohr effect	An increased H^+ concentration (lower pH) or increased Pco_2 causes a shift to the right; decreases in H^+ concentration or Pco_2 cause shifts to the left. This behavior is called the Bohr effect. It promotes O_2 uptake in the lung and O_2 release in the tissues
Temperature	At lower temperatures the curve shifts to the left (at high temperatures, it shifts to the right). In humans this is normally of no great significance
2,3-DPG	Elevated concentrations of 2,3-diphosphoglycerate (2,3-DPG) in erythrocytes cause a shift to the right
CO	Carbon monoxide (CO) has a high affinity for Hb. It binds to form carboxyhemoglobin (HbCO), which cannot bind O_2. A Pco of a few mmHg produces >90% HbCO, hence it is extremely toxic
Fetal Hb	HbF has a somewhat higher affinity for O_2 than Hb (curve shifted to the left). Furthermore, the Hb concentration is higher in fetal blood. These two factors promote placental gas exchange
Myoglobin	Structure similar to one subunit of Hb, hence each molecule can bind only one O_2 molecule. But it binds with high affinity. Hence, the saturation curve is steep and shifted to the left. Leftward shift denotes facilitation of O_2 binding; steepness denotes easy release

Blood gas data and pH values in arterial and venous blood of healthy young people at rest (modified from Thews, 1989)

	Po_2	So_2	$[O_2]$ l O_2/l blood	Pco_2	$[CO_2]$ l CO_2/l blood	pH
Arterial blood	95 mmHg (12.6 kPa)	97%	0.20 (20 vol.%)	40 mmHg (5.3 kPa)	0.48 (48 vol.%)	7.40
Venous blood	40 mmHg (5.3 kPa)	73%	0.15 (15 vol.%)	46 mmHg (6.1 kPa)	0.52 (52 vol.%)	7.37
Arteriovenous conc. difference			0.05 (5 vol.%)		0.04 (4 vol.%)	

CO_2 transport function of the blood

CO_2 dissociation curve of human blood at 37°C

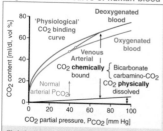

CO_2 content = amount of CO_2 that can be released by acid and vacuum, = ml CO_2/100 ml blood.

CO_2 in the blood (like O_2) is in part physically dissolved (Bunsen coefficient ca. 20 times >O_2; 1 ml blood contains 0.065 ml CO_2, 12% of total CO_2 in the blood), in part chemically bound (reversibly), as bicarbonate in erythrocyte (27%) and plasma (50%) and as carbamino-Hb (11%).

Christiansen–Douglas–Haldane effect (abbr. to Haldane effect): The CO_2 capacity of blood is inversely proportional to its O_2 saturation. This promotes CO_2 binding in tissue and CO_2 release in the lung (cf. Bohr effect for O_2).

Chemical reactions in erythrocytes during gas exchange in tissue and lung
(modified from Thews, 1989)

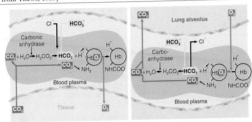

Process	Decription/comments
Hydration	After diffusion from the tissue into blood, CO_2 is hydrated (CO_2 + $H_2O \rightarrow H_2CO_3$). H_2CO_3 immediately dissociates to HCO_3^- and H^+. Due to carbonic anhydrase, CO_2 hydration occurs 10 000 times faster in erythrocytes than in plasma
Chloride shift	Also called Hamburger shift: bicarbonate (HCO_3^-) leaves the erythrocyte, in electroneutral exchange for Cl^- (via the anion exchanger or Band 3 protein)
H^+ buffering	The protons released following CO_2 hydration are buffered in blood (Hb and other buffers)
Carbamino compound (carbamate)	Formed by direct binding of CO_2 to amino groups of globin. The reaction product is called carbaminohemoglobin, abbreviated to carbhemoglobin

Acid–base status of the blood

Physicochemical background

Term	Definition/description/comments
Acids and bases	Acids release H^+ ions in solution (proton donors), bases bind them (proton acceptors). Thus in the dissociation reaction $H_A \rightleftharpoons H^+ + A^-$, HA is the acid, A^- the conjugate base (A is common to A^- and HA)
	The equilibrium follows the law of mass action (see next page). With a strong acid, e.g. HCl, the equilibrium shifts far to the right (almost complete dissociation); with a weak acid, dissociation is incomplete (see below)
pH definition	The pH is defined as the negative decimal logarithm of the molar H^+ ion concentration, $pH = -\log[H^+]$. A pH 7 solution, with $[H^+] = 10^{-7}$ mol/l, is a neutral solution. pH > 7 denotes alkaline solutions. pH < 7 denotes acid solutions
pH measurement	Indicators (color change at specific pH); electrometry with a (H^+-sensitive) glass or liquid membrane electrode

Properties of buffer systems

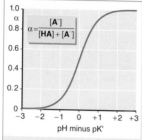

$$\alpha = \frac{[A]}{[HA] + [A]}$$

Dissociation of a weak acid follows the mass action law (where K′ is the dissociation constant, [ ] denotes molar concentration):

$$K' = \frac{[H^+]/[A^-]}{[HA]}.$$

When $[H^+]$ increases, $[HA]$ must also increase. Thus some of the added free H^+ ions are accepted by the base A^- and hence removed from the solution. Hence, the pH change is less than expected from the added H^+. The opposite occurs when H^+ ions are removed. This is called buffering.

By rearranging and taking logs of the above equation, we have

$$-\log[H^+] = -\log K' - \log\frac{[HA]}{[A^-]}, \quad \text{or} \quad pH = pK' + \log\frac{[A^-]}{[HA]} = pK' + \log\frac{[base]}{[acid]}$$

This expression of the mass action law for a buffer system is called the Henderson–Hasselbalch equation, in which pK′ is a constant characteristic of the buffer system, namely the (pH) value at which acid and conjugate base are present in equivalent concentrations (log 1/1 = 0). Further rearrangement gives

$$pH = pK' + \log\frac{\alpha}{1-\alpha} \quad \text{with the degree of dissociation } \alpha = \frac{[A^-]}{[HA] + [A^-]}$$

The relationship between α and pH of a buffer solution is shown in the figure.

Buffer capacity (buffering power): The amount of acid or base that must be added to 1 liter of a buffer solution consisting of a weak acid and its conjugate base in order to change the pH by 1 unit. The buffer capacity is greatest when pH = pK′, i.e. pH − pK′ = 0, $\alpha = 0.5$ (see Fig.) Every buffer system operates effectively only 2 pH units above or below its pK′ value, i.e. from pK′ − 2 to pK′ + 2.

Blood pH

Term	Definition/description/comments
Blood (arterial) pH	Mean 7.4, range at 37°C from 7.38 to 7.42. Applies, strictly, only to blood plasma, where it is measured. In erythrocytes the pH is 7.2–7.3
Maintenance of constant blood pH	(1) By buffer systems in the blood, see below; (2) by alveolar ventilation: excretion of the 'volatile' acid CO_2 in the expired air; (3) by H^+ excretion in the kidney: excretion of the 'non-volatile' acids of metabolism (ca. 50–60 mmol H^+ ions/day, mostly as NH_4^+ and $H_2PO_4^-$). (2) and (3) prevent H^+ accumulation in body fluids
Acidosis and alkalosis	A reduction of blood pH < 7.38 is called acidosis, elevation >7.42 is alkalosis. The lowest blood pH compatible with life is ca. 6.8. If the acid–base disturbance is due to changes in the loss of CO_2 (hypoventilation or hyperventilation), it is called respiratory (acidosis or alkalosis). If it is due to changes in the balance of non-volatile acid, it is called metabolic (acidosis or alkalosis)

Buffer systems in the blood

System	Description/comments
Bicarbonate buffer	Weak acid: carbonic acid, H_2CO_3, conjugate base: bicarbonate, HCO_3^-, pK' = 6.1. Although its pK' is far from the blood pH, an important buffer because a high, constant bicarbonate concentration (24 mmol/l) is normally regulated by renal excretion of H^+ and because P_{CO_2} is regulated by alveolar ventilation
Phosphate buffer	Weak acid: primary phosphate, $H_2PO_4^-$, conjugate base: secondary phosphate, $H_2PO_4^{2-}$, pK = 6.8. Phosphate concentrations in blood are small, hence buffer role is minor (but important in urine)
Protein buffers	Important here are the ionizable side groups of the amino acids, especially the imidazole ring of histidine. Buffers include the plasma proteins (mainly albumin) and Hb (chief protein buffer), the latter especially in deoxygenated form, because it is less protonated than oxygenated Hb and hence better able to buffer H^+ ions

Buffer base (BB) and base excess (BE) values

Term	Definition/description/comments
Total buffer base	Sum of all anions with buffer action in the blood. Amounts to ca. 48 mmol/l, proteins and bicarbonate each about half (values above). Stays largely constant when P_{CO_2} changes, because an increase in bicarbonate concentration is compensated by a decrease in protein dissociation and vice versa
Base excess (BE)	BE is determined by measuring how much acid is needed (in mmol/l) to bring blood at 37°C with an arterial P_{CO_2} of 40 mmHg to the pH 7.4. For instance, if 5 mmol/l acid is needed, the BE value is +5 mmol/l. Normally BE = zero; only in case of excess of either non-volatile acids or bases (metabolic acidosis or alkalosis) is there a base deficit (negative BE) or base excess (positive BE), respectively (see Fig.)

CO_2 dissociation curve of the blood as a function of BE (see also Fig. p. 206)
(modified from Thews, 1989)

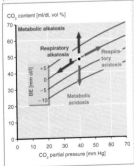

In respiratory alkalosis (due to hyperventilation) or respiratory acidosis (hypoventilation) the CO_2 dissociation curve is unchanged (for pH change see next Fig.). In metabolic alkalosis BE increases and the CO_2 dissociation curve shifts in the direction of increasing CO_2 content; in metabolic acidosis there is a base deficit (negative BE) and shift of the CO_2 dissociation curve toward decreasing CO_2 content. Hence one can differentiate between respiratory and metabolic disturbances by measuring P_{CO_2} and BE.

Bicarbonate concentration in arterial blood plasma in primary, uncompensated disturbances of acid–base balance, represented by the pH–bicarbonate diagram (based on J. Piiper, 1975)

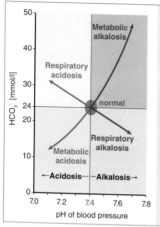

This, the preceding and the following figures represent the same situations from different viewpoints. Example: respiratory acidosis is characterized by an elevated arterial P_{CO_2} (hypercapnia, Greek *kapnos*: smoke) (see preceding figure), and at the same time the plasma bicarbonate increases (left) because of buffering of H_2CO_3 by non-bicarbonate buffers. But this does not suffice, so that the pH falls (left and below). The acidosis is partially compensated by renal excretion of protons (see below). The reverse applies to hypocapnia (respiratory alkalosis) produced by hyperventilation. The compensation of metabolic disturbances is respiratory (see below).

Disturbances of acid–base balance and their compensation in the blood, illustrated by the P_{CO_2}–pH diagram (based on J. Piiper, 1975)

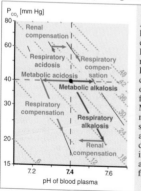

Abscissa: blood pH. Ordinate: arterial P_{CO_2}, logarithmic scale. The 4 primary disturbances are departures from the normal blood pH (7.4) and P_{CO_2} (40 mmHg). The compensatory mechanisms restore the blood pH to values close to normal. Due to buffering in the blood, the lines for the changes associated with respiratory disturbances or compensations have a steeper slope than those for constant bicarbonate content (thin lines). Mixed forms (combined respiratory and non-respiratory acidosis or alkalosis) also exist (not shown). Metabolic acidosis is common (e.g. lactic acidosis during heavy exercise, ketoacidosis in diabetes mellitus, loss of alkaline intestinal fluid in diarrhea, renal insufficiency). Metabolic alkalosis is less common (e.g. vomiting of gastric fluid, use of diuretics).

Diagnosis of acid–base status

Nomogram of the acid–base status of the blood at 37°C (after Siggaard-Andersen, 1963)

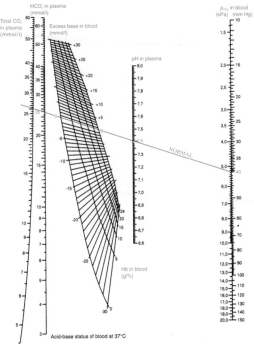

Acid-base status of blood at 37°C

Interpretation/definition of acid–base disturbances (cf. also definition of acidosis [pH below 7.38] and alkalosis [pH above 7.42] p. 208). Hatched rectangles: normal range

P_{ACO_2} ＼ HCO_3^-	Below 21 mmol/l	21–29 mmol/l	Above 29 mmol/l
Above 44 mmHg	Combined metab. and resp. acidosis	Respiratory acidosis	Metabolic alkalosis and resp. acidosis
34–44 mmHg	Metabolic acidosis	Normal	Metabolic alkalosis
Below 34 mmHg	Metabolic acidosis and resp. alkalosis	Respiratory alkalosis	Combined metab. and resp. alkalosis

Tissue O₂ consumption and tissue metabolism

Features of aerobic and anaerobic energy-providing mechanisms

Term	Definition/features/comments
Aerobic	Oxidative metabolism of fats, proteins and carbohydrates in mitochondria. End-products always include CO_2 and H_2O, for protein breakdown also N-containing substances, mainly urea
Anaerobic	Glycolysis. End-product lactate (still very high-energy). 1 mol glucose gives 208 kJ = 50 kcal; oxidative breakdown gives ca. 15 times as much energy (2883 kJ = 689 kcal). Hence anaerobic metabolism is only a short-term solution, for start-up and emergencies. High lactate levels may cause metabolic acidosis
O₂ supply	Product of arterial O_2 concentration and blood flow (selected values on p. 176). Arterial O_2 concentration is the same everywhere, so O_2 supply depends entirely on blood flow (very high, e.g. in renal cortex, cerebral cortex; very low, e.g. in resting skeletal muscle)
O₂ utilization	Ratio of the O_2 consumption of an organ to its O_2 supply. At rest, e.g. in cerebral cortex, myocardium, skeletal muscle, ca. 40–60%. Maximal utilization at peak performance ca. 90%. In contrast, very low in kidney and spleen
Q₁₀	O_2 consumption rises by factor of 2–3 per 10°C increase in the temperature range 20–40°C. Artificial hypothermia is used to conserve energy during surgery, e.g. open-heart operations

Gas diffusion in tissue: partial pressures in the various parts of the circulatory system at rest (after Thews, 1963)

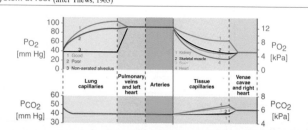

Gas exchange in the tissues, as in the lungs, occurs by diffusion (governed by Fick's first law of diffusion). Factors are exchange area, diffusion distance, diffusion resistance and partial pressure differences as driving forces. O_2 molecules move out of erythrocyte and plasma into tissues. Transport to erythrocyte surface is assisted by diffusion of the oxygenated hemoglobin within the erythrocyte: facilitated O_2 diffusion. In skeletal muscle diffusion of the oxygenated myoglobin has a similar influence on O_2 transport.

Models of respiratory gas exchange in tissues, significance of capillarization
(figure modified from Grote, 1989)

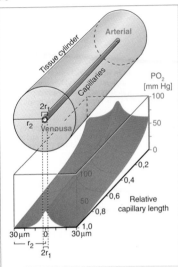

Models are used to study the conditions for gas exchange in tissues; best known is Krogh's tissue cylinder. Applies strictly only to capillaries in parallel through which blood flows in the same direction. The figure reflects situation in cerebral cortex. There also exist cone models (blood flows through capillaries in opposite directions), cubic tissue column (four parallel capillaries with blood flow in different directions) and complicated networks.

Capillary density in tissue determines the exchange area for respiratory gases and the diffusion distances. It varies greatly from organ to organ, e.g. mean capillary distance in myocardium 25 µm, in cerebral cortex 40 µm, in skeletal muscle 80 µm. Exchange area and diffusion distance can be modified by arteriolar vasoconstriction or vasodilation.

In brain tissue the O_2 partial pressure of the blood falls from 90 mmHg to ca. 28 mmHg during passage through a capillary. Between the blood and the periphery of the supplied tissue cylinder, the O_2 partial pressure difference is ca. 26 mmHg. Hence in the most poorly supplied cells in the outer venous region, the O_2 partial pressure is expected to be only about 2 mmHg.

Critical O_2 partial pressure in the mitochondria

Most important criterion for evaluating an organ's O_2 supply is the cellular O_2 partial pressure. For normal oxidative metabolism it must be at least 0.1–1 mmHg (13.3–133.3 Pa) in the vicinity of the mitochondria. Can be measured with platinum microelectrodes by the polarographic method; otherwise estimated by means of the above tissue model.

Special cases: myocardium and skeletal muscle. Here myoglobin serves as a short-term O_2 storage site ('O_2 buffer') and intracellular O_2 transporter (facilitated diffusion, see above). As a result, partial pressure differences between the cells are reduced and O_2 partial pressures are kept nearly constant when a load is applied, despite increased O_2 consumption.

Regulation of O$_2$ supply

Regulation of blood flow in specific organs: see pp. 176, 190 (under normal conditions O$_2$ supply depends entirely on blood flow, see p. 176)

The 3 main causes of deficient O$_2$ supply

Cause	Description/comments
Arterial hypoxia	Results from alveolar hypoventilation; in the arterial blood the O$_2$ partial pressure is reduced (hypoxia) and so is the O$_2$ concentration (hypoxemia), with accompanying respiratory acidosis (hypercapnia, see p. 210). At high altitudes, arterial hypoxia is accompanied by respiratory alkalosis (hypocapnia)
Anemia	A reduced O$_2$ transport capacity of the blood can be caused by (1) deficient hemoglobin production, (2) hemoglobin loss (1 and 2: anemia in the narrow sense), (3) methemoglobin formation and (4) CO poisoning ('functional anemia')
Ischemia	When organ blood flow is restricted, more O$_2$ is extracted from the blood during passage through the capillaries, so that the arteriovenous O$_2$ concentration difference increases. The result is venous hypoxia

Influence of anemia on the O$_2$ partial pressure changes in capillary blood (modified from Grote, 1989)

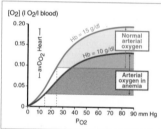

The graph shows the conditions in the myocardium at rest, comparing normal Hb content (15 g/dl) and low Hb content (10 g/dl). The ordinate shows O$_2$ concentration in ml O$_2$ per ml blood, the abscissa shows O$_2$ partial pressure. Note that under these conditions, with no change in amount of O$_2$ withdrawn by the tissue during passage through the capillaries, the blood O$_2$ concentration can become very low (venous hypoxia with danger of falling below the critical O$_2$ partial pressure for the mitochondria, see above).

Effects of O₂ deficiency

Changes in cell metabolism during and after acute ischemic hypoxia
(from Grote, 1989)

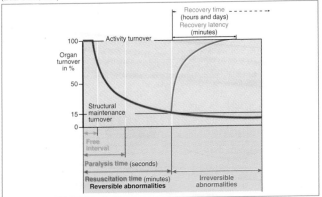

Basic terms used to describe acute tissue hypoxia, e.g. due to complete
ischemia or marked arterial hypoxia (see also above Fig.)

Term	Definition/comments
Latent period	Time during which cell function is unimpaired. As a rule it is a few seconds, e.g. ca. 4 s in the brain
Deterioration time	Time from the onset of tissue hypoxia to complete loss of organ function. Also usually brief, e.g. 8–12 s in brain
Resuscitation time	Time after the onset of tissue anoxia, during which complete resuscitation of the whole organ is still possible. Varies widely from organ to organ, e.g. active heart 3–4 min, brain at 37°C 8–10 min, kidney and liver 3–4 h, organism as a whole only 4 min
Recovery latency	Time from the end of hypoxia to the (in some cases initially incomplete) restoration of organ function
Recovery time	Time from the end of hypoxia to complete restoration of organ function. Can be as long as days, e.g. 15 min after brain ischemia lasting only 1 min

O₂ therapy can cause O₂ poisoning

Isobaric and hyperbaric O₂ therapy to improve the O₂ supply conditions in the tissues (e.g. in cases of myocardial infarction, asthma, CO poisoning) must be made as brief as possible, because a great increase in cellular O₂ partial pressure (hypoxia) inhibits enzymes involved in tissue metabolism (symptoms of such O₂ poisoning include, e.g. vertigo, seizures, long-term damage of lung and retina)

Parameters of metabolism

Units: 1000 cal = 1 kcal = 4190 J = 4.19 kJ ~ 0.0042 MJ and
       1000 J = 1 kJ = 239 cal = 0.239 kcal (for details see p. 2)

Levels of cellular metabolic activity

Level	Definition/description/comments
Active level	Metabolic rate of an active cell, depends on the degree of activity at any time
Readiness level	Rate at which a momentarily inactive cell must metabolize in order to maintain its capacity for immediate, unrestricted function
Maintenance level	Essential, minimal metabolic rate necessary to preserve cell structure; with any lower rate, the cell would die

Metabolic parameters of the whole organism

Term	Definition/description/comments
Resting rate	Metabolic rate at mental and physical rest. Note: some activities continue at rest, e.g. heartbeat, breathing, brain function. Hence resting metabolism > sum of readiness metabolism of all cells
Basal metabolic rate (BMR)	Resting metabolism under standard conditions: (1) morning, (2) resting (lying down), (3) fasting, (4) neutral ambient temperature, normal body temperature. Practically all departures from these conditions increase metabolic rate
Working rate	Metabolic rate during work, hence BMR plus energy required for work (total metabolic energy). Minimally 'leisure rate' (energy needs of a 'desk worker' doing no physical labor; see Table on next page). Metabolic rate also rises during mental work due to an increase in muscle tone
Efficiency	The efficiency (η) of metabolism is the ratio of external work performed to the total metabolic energy (gross efficiency) or to the energy added for work (net efficiency). The latter is between 10% and 25% for exercise

Basic metabolic rate (BMR) of humans as a function of age and sex, and the contributions of various organs (after Boothby *et al.*, 1936)

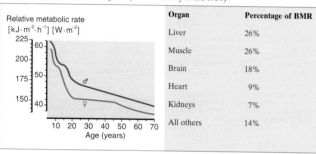

Organ	Percentage of BMR
Liver	26%
Muscle	26%
Brain	18%
Heart	9%
Kidneys	7%
All others	14%

General principles

Energy metabolism and associated O_2 uptake under typical conditions
(from Ulmer, 1989a)

Condition	Metabolic rate			O_2 uptake (ml/min)
		(kJ/d)	(W)	
Basal metabolic rate (BMR), 70 kg body weight	Woman:	6 300	76	215
	Man:	7 100	85	245
BMR plus 'leisure' increment (moderate activity)	Woman:	8 400	100	275
	Man:	9 600	115	330
Total metabolic rate for maximal exercise, long-term (years)	Woman:	15 500	186	535
	Man:	20 100	240	690

For short periods during athletic activities, metabolic rates considerably higher than those shown above can be reached, e.g. ca. 550 W in normal cycling, 500–1000 W in playing tennis and 1500 W in running a marathon.

Assessment of metabolic rate by measurement of O_2 consumption (indirect method)

Respiratory quotient (RQ) and energy equivalents (caloric equivalent, in kJ/l O_2 consumption) for the oxidation of various foodstuffs

	Carbohydrates	Fats	Proteins
RQ	1.00	0.70	0.81
kJ/l O_2	21.1	19.6	18.8

RQ is the ratio of CO_2 release to O_2 uptake during respiration. When glucose is oxidized just as much CO_2 is given off as O_2 is consumed, hence the RQ is 1. If only fat or only protein is oxidized, more O_2 is consumed than CO_2 is given off, producing the above quotients.

Relationship between energy equivalent and RQ, neglecting the protein contribution (15%) to total metabolism. Under BMR conditions, RQ is usually 0.82. On the whole, the margin of error is low (±4% of 20.2)

RQ	1.0	0.9	0.82	0.8	0.7
kJ/l O_2	21.1	20.6	20.2	20.1	19.6

The proportion of protein in the total amount of food oxidized can be determined by measuring the amount of nitrogen (mainly urea) excreted in the urine and multiplying that value by 6.25, because protein has an average N content of 16%. The proportion of protein under normal conditions is about 15% of the total metabolic rate, and hence can usually be neglected for practical purposes. The proportions of fats and carbohydrates and associated energy equivalents can therefore be estimated with sufficient accuracy from the RQ.

Changes in RQ unrelated to dietary foodstuff proportions

During hyperventilation, the amount of CO_2 eliminated from the body rises. The result is a temporary distinct increase in RQ, in some cases to 1.4. Excess carbohydrate in the diet is converted to fat, releasing O_2; hence less O_2 is taken up and RQ rises (e.g. force-feeding raises RQ in geese to 1.38, in pigs to 1.58). Conversely, fasting people and diabetics have lower RQ values, down to 0.6 (because when glucose metabolism is diminished, the rate of fat and protein conversion rises).

Calorimetric measurement of metabolic rate (direct method)

Calorimeters are devices designed to measure directly and continuously the heat generated by organisms. They are needed only for special research purposes. Calorimeters have been used, for instance, to demonstrate that the law of conservation of energy also applies to living organisms. The validity of indirect measurement methods has also been established by direct calorimetry.

Measurement of O_2 consumption (i.e. O_2 uptake)

Measurement of O_2 uptake in a closed spirometer

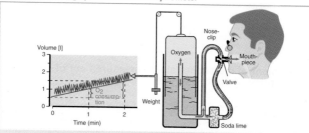

Pure O_2 must be used. Advantage: no need to measure O_2 concentration. CO_2 is absorbed with soda lime. Disadvantage: impossible to measure RQ. O_2 uptake must be converted to STPD conditions (see p. 193) by calculation or reference to tables. (An air-filled spirometer would produce an 'altitude experiment': rapid decrease in O_2 partial pressure with constant low CO_2 partial pressure, which soon impairs consciousness with no subjective shortness of breath or other warning signals. Omission of soda lime and filling with air would produce a 'rebreathing experiment': increased CO_2 partial pressure would rapidly cause hypercapnia, intensify breathing and cause subjective shortness of breath.)

Measurement of O_2 uptake in open respiration systems

Method	Description/comments
Douglas bag	Nose closed by a clamp. Fresh air inhaled through a valve mouthpiece. Expiration into the (portable) bag. After the measurement period the bag is emptied through a gas meter, and CO_2 and O_2 concentrations are measured
Principle of the aliquot portion	Nose closed by a clamp. Fresh air inhaled through a valve mouthpiece. Volume of expired air measured with gas meter or pneumotachograph. Under the control of the gas meter, small samples containing all components of the expired air in the correct proportions ('aliquots') are pumped out for gas analysis
Principle of constant suction	Nose closed by a clamp. Breathing by way of valveless mouthpiece. Pump sucks a mixture of air from the inflow opening and expired air through the gas meter, and at the same time a subfraction is taken for continuous gas analysis, from which O_2 uptake and CO_2 release can be calculated

Heat production and body temperature in humans

Heat production and maintenance of body temperature

Form	Description/comments
Cellular metabolism	Each form of metabolism described in the preceding chapter is associated with heat production, according to the laws of thermodynamics. This is the main source of human body heat
Exercise	Variable form of heat production (see table at top of p. 217). Body movement generates heat, often more than needed to keep body temperature constant. Can be voluntarily employed to produce heat
Shivering	Muscle activity used solely to generate heat. Before perceptible shivering, muscle tone is increased. In adults, shivering is the most important involuntary mechanism for producing additional body heat
Non-shivering thermogenesis	Heat production by increased metabolism in tissues other than skeletal muscle, mainly brown adipose tissue. It is important in newborns, insignificant or absent in adults

To keep body temperature constant, a state of thermal balance must be maintained, i.e. heat production must equal heat loss. This is achieved by thermoregulation (see below). When heat loss exceeds heat production, the result is hypothermia which can be lethal. Insufficient heat loss relative to heat production (e.g. too-warm surroundings) leads to hyperthermia, which can also be lethal. The thermoneutral zone is the range of ambient temperatures in which the body temperature can be kept constant without additional heat production or without production of sweat.

Isotherms in the human body in cool (20°C, A) and warm (35°C, B) surroundings (after Aschoff and Wever, 1958)

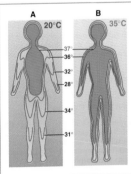

A   B
20°C   35°C

—37°
—36°
—32°
—28°

—34°

—31°

In a simplified view, a homeothermic core of the body can be distinguished from a poikilothermic shell.

The temperature in the body shell depends on ambient temperature and on the degree of heat production.

In the limbs there is both an axial and a radial temperature gradient. In cold surroundings, the 37°C isotherm is withdrawn into the body.

The core temperature is also not absolutely equal everywhere, e.g. in the brain there is a radial gradient of ca. 1°C from center to cortex. The highest temperature is in the rectum (commonly used for clinical measurement), the sublingual temperature (also used) is 0.2–0.5°C lower.

Periodic variation in the core temperature of the body (modified from T.H. Schmidt as cited in Brück, 1989)

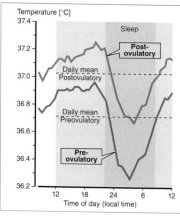

The diurnal rhythm of body temperature (minimum: early morning; maximum: late afternoon) and the sleep–wake cycle are the best-known forms of circadian periodicity (see p. 115).

Body temperature varies not only in a 24 h cycle but also over longer periods, e.g. the postovulatory elevation of temperature in women during the menstrual cycle (see p. 267).

Exercise raises core temperature (e.g. in marathon runners to 39–40°C). For other reactions that adjust the body to dynamic and static work see pp. 228–230.

Heat loss

Internal heat flow

Term	Definition/description/comments
Internal heat flow (H_{int})	The amount of heat transported from the interior to the surface of the body is proportional to the skin surface area A and to the difference between core temperature T_c and mean skin temperature $\bar{T}_s$, hence in watts $H_{int} = C(T_c - \bar{T}_s)A$ [W]
Thermal conductance (C)	The proportionality factor in the above equation depends on the rate of blood flow through the skin and through the limbs; its reciprocal $1/C = I_t$ is called the thermal resistance or thermal insulation of the body shell
Convection	Heat transport by flowing blood; is the main mechanism of internal heat transport
Conduction	Heat transport through the tissues is less important
Countercurrent principle	Because the arteries and deep veins in the limbs lie in parallel, the blood flowing back through the veins 'extracts' heat from the arteries (countercurrent heat exchange), so that the peripheral regions receive precooled blood. In warm surroundings, because of dilation of superficial skin veins, the peripheral regions become warmer (the factor C changes accordingly)

External heat flow

Term	Definition/description/comments
External heat flow (H_{ext})	Heat transport from the body surface to the surroundings has the components listed in the next four rows of this table: $H_{ext} = H_k + H_c + H_r + H_e$
Heat transfer by conduction (H_k)	Occurs only when the body contacts a solid substrate (e.g. sitting or lying on wood, metal, etc.). On the whole, conduction is quantitatively unimportant, practically negligible during intense exercise. Clothing allows only conductive heat loss, which is the basis of its insulating property
Convective heat transfer (H_c)	Occurs when the air next to the skin is warmed and therefore rises: natural or free convection. Can be considerably increased by externally imposed air movement: forced convection. The rate of heat loss is also determined by the temperature difference between skin and air, the area of the body surface that is exposed, and the wind speed
Heat exchange by radiation (H_r)	Described by the Stefan–Boltzmann equation. H_r and H_c constitute the 'dry' heat loss; here the ambient temperature is considered to be the operative temperature, a weighted mean of the air and radiant temperatures
Evaporative heat transfer (H_e)	At physical rest in a neutral ambient temperature, this basically occurs by water transport through the skin and the mucosa of the respiratory tract, followed by evaporation: insensible perspiration, accounting for ca. 20% of total heat loss. During exercise and/or at high ambient temperatures, most heat transfer occurs by sweating (see Fig. below). At ambient temperatures above body temperature, evaporation is the only way for heat to be dissipated

Heat production and heat loss at rest and during work (after Dubois, 1937)

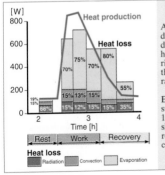

At rest, heat is lost mainly by radiation; during exercise, evaporative heat loss predominates. Conductive heat loss is ignored here. The core temperature of the body rises during exercise (see above), whereas the skin temperature falls due to the evaporation of sweat.

Evaporative heat loss is possible even in surroundings with a relative humidity of 100%, as long as the vapor pressure on the skin surface is greater than that in the surroundings, i.e., when skin temperature exceeds ambient temperature (see p. 194).

Environmental factors and thermal comfort

Thermal comfort is the state in which, with an average rate of peripheral blood flow, neither shivering nor sweating is necessary; i.e. the organism is not under thermal stress. Four factors are important: (1) air temperature, (2) humidity (water vapor pressure), (3) radiation temperature and (4) wind speed (wind chill factor). Some of these factors are 'interchangeable' in the sense that, e.g. a too-low radiation temperature (room with cold walls) can be compensated by a higher air temperature.

The comfort temperature in water is distinctly higher than that in air, because water withdraws much heat from the body by convection and conduction. At complete rest this temperature is 35–36°C. When the body is moving in water (forced convection), the turbulent water currents extract so much heat from the skin that hypothermia is unavoidable at water temperatures of 10°C or less.

Guidelines for neutral and comfort temperatures in adults (after Wenzel and Piekarski, 1982)

Surroundings	Other conditions	Clothing	Temperature range
Air	No wind, 40–50% humidity, physical rest, neutral radiation conditions	Normal street clothing	20–22°C
Water	At rest (bathtub) Swimming, 0.4 m/s	Nude, swimsuit Nude, swimsuit	35.5–36°C 28°C

Control of body temperature

Block diagram of the control system for thermoregulation
(modified from Brück, 1989)

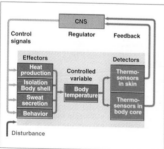

Body temperature is kept constant by negative feedback regulation. The control system and its variables and the neural elements involved are shown in the figure

Thermoregulatory behavior includes putting on and taking off clothing, staying in the shade and so on.

The thermoregulatory center in the posterior hypothalamus (see p. 224) is connected to the motor centers in the brainstem by way of the central shivering pathway.

Heat loss

Term	Description/comments
Peripheral blood flow	The peripheral areas are the fingers, toes, ears, lips and nose. In the absence of vasoconstrictor tone the arterioles and arterio-venous anastomoses dilate maximally, causing greatly increased blood flow and hence a great increase in convective heat transport (the opposite occurs when vasoconstrictor tone is high)
Blood flow through the skin of the trunk	In the trunk and proximal parts of the limbs there is another means of vasodilation (other than decreased vasoconstrictor tone), either active neuronal vasodilation or the action of a chemical mediator (e.g. bradykinin), which is released with the sweat (the latter is more likely than nerve-mediated vasodilation, which does not exist in man)
Emotional sweating	Severe mental tension can lead to the paradoxical situation of cutaneous vasoconstriction in the hands and feet accompanied by sweating at the palmar and plantar surfaces ('cold sweat'). In contrast, thermal sweating is accompanied, as would be expected, by vasodilation
Lewis reaction	Cold vasodilation: exposure to extreme cold can cause maximal vasoconstriction to be followed by sudden vasodilation, repeated periodically if exposure is prolonged; may be due to a direct action of the cold on arteriolar smooth muscle. The vasodilation can cause overcooling, e.g. in accidental cold-water immersion

Neural control of thermoregulatory effector elements, including transmitters, receptors and antagonists (modified from Brück, 1989)

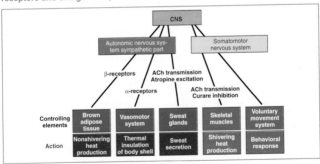

Structures involved in thermoregulation and their function (for the effectors in thermoregulation see figure on p. 223)

Structure	Description/function/comments
Cutaneous temperature receptors	Function of cutaneous 'cold and warm sensors' described on p. 37. They 'inform' the thermoregulatory system (and consciousness) about the temperature of the body shell and its changes. Together with the internal thermosensors, they provide the thermoregulatory center with an accurate picture of the thermal status of the core and shell of the body
Internal temperature receptors	Thermosensitive neurons ('warm neurons') are located in: 1. The anterior hypothalamus (preoptic region, high temperature sensitivity) 2. The lower brainstem (midbrain and medulla oblongata, slight temperature sensitivity) 3. The spinal cord (exact location of the warm neurons unknown, but high temperature sensitivity) The warm neurons serve as internal thermosensors: local warming or cooling of warm neurons elicits heat-loss responses (e.g. sweating) or enhances heat production (e.g. shivering), respectively Internal thermosensors are thought to exist outside the CNS, e.g. at the dorsal wall of the abdominal cavity and in the musculature. Their existence is not certain
Thermoregulatory center	Located in the posterior hypothalamus, which itself has no thermosensitivity. It 'integrates' information from the thermosensors and sends appropriate control signals to the effectors for thermoregulation Unlike those of the anterior hypothalamus, the neurons of the posterior hypothalamus are thermoresponsive and not thermosensitive. The goal of thermoregulation is to keep body temperature as close as possible to the set point (which varies periodically, see above)
Afferent and efferent neural pathways	The signals conveyed by thermosensors in the trunk and limbs pass over reticular branches of the spinothalamic tract, and those from the head pass through the caudal trigeminal nucleus to the thermoregulatory center in the posterior hypothalamus. The warm neurons of the spinal cord are connected to the latter by the anterolateral funiculus Efferent output to the motor system is mediated by the central shivering pathway mentioned above. Vasomotor outputs are sent from the posterior hypothalamus by way of the medial forebrain bundle to the autonomic nuclei in the spinal cord

Ontogenetic and adaptive changes in thermoregulation

Thermoregulation in premature infants and newborns

In principle, it operates as in adults. Special features include:

1. Instead of shivering in cold conditions, non-shivering thermogenesis is the main mechanism (not realized for a long time, because it is detectable only by special procedures).

2. The ratio of body surface to body volume is about five times as large as in adults. Body shell and fat insulation are thin, hence favoring high rate of heat loss. Therefore, body temperature can be kept constant only in relatively high ambient temperatures (32–34°C when unclothed with minimal metabolic rate, barely 28°C during maximal heat production, compared with 0°C for adults).

Physiological adaptation to extreme climates (acclimation, in addition to behavioral mechanisms involving clothing, housing, heating, etc.)

Term	Definition/description/comments
Heat adaptation	In the tropics, in the desert, in some work environments. The following adjustments occur, because of resetting of the set point:
	1. Rate of sweat production increases by a factor between one and two
	2. Threshold for sweating is shifted to lower core and shell temperatures
	3. Sensation of thirst increases
	In hot, damp climates, a decrease in the rate of sweat secretion frequently takes place after a period of profuse sweating (hyperhidiosis). Mechanism unknown. Reduces the thermally useless dripping of sweat
Tolerance adaptation to heat	Found only in inhabitants of the tropics who avoid hard physical labor as much as possible: the sweating threshold is shifted to higher body temperatures
Cold adaptation	During periodic exposure to cold (e.g. pearl divers) there is a reduction in shivering threshold and other responses to cold, so that moderate hypothermia develops: tolerance adaptation to cold
	A different reaction is observed in response to prolonged cold stress (Eskimos, Patagonian Indians): basal metabolism is increased by 25–50%: metabolic adaptation

Pathophysiological aspects of thermoregulation

Fever

The core temperature is regulated at a higher level by a shift of the set point. With fever, thermogenesis is greatly increased by shivering, and at the same time peripheral vasoconstriction minimizes heat loss.

Conversely, during recovery from fever (set point returned to the normal level) there is extreme sweat production and vasodilation.

Substances that produce fever are called pyrogens. Exogenous pyrogens (e.g. bacterial endotoxins) induce the formation of endogenous pyrogens. For details see pathophysiology textbooks.

It is still debated whether fever helps the body to fight infection or whether it is a harmful side-effect of immune responses.

Conditions in which the tolerance limits of the control system are exceeded

Term	Description/comments
Hyperthermia	Increased body temperature when the heat-loss mechanisms are overloaded. Elicits a more unpleasant subjective sensation than equally high fever. Briefly tolerable up to 41°C (e.g. marathon runners, see above). When persistent, leads to heat stroke: severe, often rapidly lethal brain damage, mainly due to failure of thermoregulation, with cessation of sweat secretion, onset of 'paradoxical' shivering; symptoms include delirium and seizures
Heat syncope	Special form of orthostatic collapse (p. 183): extreme vasodilation under heat stress during quiet standing causes large amounts of blood to collect in the veins of lower extremities and abdomen, resulting in low venous return and a fall in cardiac output and blood pressure with eventual loss of consciousness. Relatively harmless
Malignant hyperthermia	Excessive increase in metabolic rate and hence extreme thermogenesis in skeletal musculature under general anesthesia. If untreated can be lethal. Apparently genetically determined
Hypothermia	A decrease in body temperature as soon as heat losses can no longer be compensated by heat production (e.g. during accidental immersion in cold water). Associated with respiratory and metabolic acidosis. Death occurs at core temperatures of 26–28°C due to ventricular fibrillation

Intentionally induced hypothermia can be used therapeutically, the thermoregulatory processes having been inactivated (e.g. by general anesthesia) at the hypothalamic level |

Units for the measurement of performance (work per unit time):
$1 W = 1 J/s \sim 0.1 m \, kp/s$ (for details see Tables pp. 1, 2)

Basic terminology of work physiology

Term	Definition/description/comments
Load	The physical or psychological task externally imposed on the individual. Physical loads are easily measurable, psychological ones only with difficulty
Performance	The individual's performance in carrying out the task can also be either physical (involving muscular effort) or psychological (mental, emotional). In physical performance, dynamic work is distinguished from static work. The former is work in the physical sense (force times distance) and hence is easily measured. Static work is postural work (isometric muscle contraction). Physically no work is done, but the strain can be measured
Strain	The degree of strain experienced in producing a particular performance depends not only on the load but also on the individual's performance capacity (which depends on talent, health and state of training as well as environmental influences) and on his or her efficiency (effectiveness of effort). Strain can be measured by indicators, of which the main ones are heart rate, blood pressure and O_2 uptake
Ergometry	Procedure to determine physical performance capacity (bicycle ergometer, treadmill ergometer, also knee bends or stair climbing for simple tests). Efficiency of work on ergometers is 20–25%
Fatigue	A decreased performance capacity resulting from hard work. Muscular or physical fatigue is based on changes in skeletal muscles (depleted energy stores, accumulation of lactic acid). Strenuous mental or monotonous work causes central or psychological fatigue (impaired thinking, aversion to the work)
Recovery	Begins when performance is stopped or reduced. Performance capacity increases again. Recovery is best at the beginning of a pause, hence many short pauses are better than a few long ones. Recovery also occurs when physical performance is below the long-term performance limit (see below)
Exhaustion	Acute: rapid decrease in performance capacity during physical or psychological performance at levels above the long-term limit, when there is insufficient opportunity for recovery or it is not provided soon enough. Massive metabolic acidosis. Chronic: physical collapse after prolonged, extreme strain without adequate pauses for recovery
Overloading	The occurrence of damage due to an acutely or chronically excessive workload (e.g. bone fractures, torn muscles and tendons)

Adjustments to physical work

Muscle metabolism and blood flow through muscles during dynamic work
(figure after Keul *et al.*, 1969)

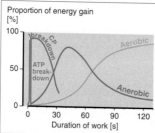

Proportion of energy gain [%]

Duration of work [s]

The ATP reserves are used up within a few seconds and those of creatine phosphate (CP) soon thereafter. Anaerobic glycolysis reaches a maximum after about 45 s, while aerobic metabolism needs ca. 2 min to become fully operative.

During heavy muscular work some of the energy is always obtained anaerobically.

Blood flow through the muscles is relatively low at first (20–40 ml/kg per min), rises within 20–30 s, depending on the level of work, to a maximum of 1.3 l/kg per min in untrained people or 1.8 l/kg per min in those with endurance training. During heavy exercise, the blood supply does not meet the demand (hence there is anaerobic metabolism [see above], accumulation of lactic acid and fatigue).

Cardiovascular function during dynamic work
(figure based on E.A. Müller, from Ulmer, 1989b)

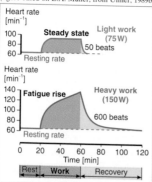

Heart rate [min⁻¹]

Time [min]

Heart rate rises during light exercise (top) over 5–10 min, to a plateau (steady state), the level of which depends on the strain. Return to the starting level after work takes 3–5 min.

Heavy exercise (bottom): Heart rate shows fatigue rise to individual maximum (followed by exhaustion, see above). Recovery takes up to several hours. Recovery pulse sum: number of pulse beats above initial level during recovery (yellow area in figure, measure of preceding strain and muscular fatigue).

Stroke volume: rises at beginning of exercise period by 20–30% and then remains constant. Thereafter, cardiac output and heart rate are practically proportional to one another. At very high heart rates, the stroke volume can decrease because the filling time (diastole) is shortened.

Systolic blood pressure: increases in proportion to performance. On average, for 100 W is maximally 200 mmHg, for 220 W maximally 220 mmHg. Diastolic blood pressure: hardly any change. Arterial mean pressure: slight to moderate increase. Low-pressure system: little change.

O$_2$ uptake and breathing during dynamic work (figure modified from Ulmer, 1989b)

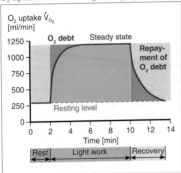

O$_2$ uptake rises during light exercise over 3–5 min after onset, to equilibrium with O$_2$ demand (steady state). Until then, the greater O$_2$ demand is met by using up myoglobin O$_2$ (pp. 205, 206) and Hb-O$_2$ (increased arteriovenous difference). An oxygen debt develops (blue area), which is paid back at the end of work by an O$_2$ uptake above the resting level (red area).

During heavy exercise no equilibrium can be established between O$_2$ uptake and demand. Like the heart rate (see above), O$_2$ uptake rises to a maximum (followed by rapid exhaustion). During dynamic work heart rate and O$_2$ uptake increase in parallel, and both are proportional to performance (given constant efficiency).

Respiratory minute volume: during light exercise rises in proportion to O$_2$ uptake (see regulation of breathing, p. 201). During heavy exercise, metabolic acidosis (increased lactic acid production) stimulates ventilation, so that ventilation increases disproportionally more than O$_2$ uptake.

Blood parameters during dynamic work

Parameter	Description/comments
Blood gases	In light exercise there is practically no change. In heavy exercise, there is a slight reduction in the arterial partial pressures of O$_2$ (at most –8% from resting level) and CO$_2$ (at most –10% from resting level). The arteriovenous O$_2$ difference increases (typically from 5 vol.% to 15 vol.% [cf. p. 176]), because of the increased O$_2$ consumption
Acid–base status	In light exercise, there is practically no change. In heavy exercise, the lactic acid produced (see above) causes metabolic acidosis, which is partially compensated by hyperventilation (see also p. 210)
Blood cells	The hematocrit rises for two reasons: increased ultrafiltration in the muscle capillaries, and release of erythrocytes from the sites of hematopoiesis. There is also exercise leukocytosis (greater release of leukocytes)
Nutrients	In light exercise, there is practically no change. In heavy exercise, there is mainly an increase in lactate concentration (see above), e.g. from 1 mmol/l to 15 mmol/l

Thermoregulation during dynamic work (see also pp. 219, 222, 224)

Excess heat loss during exercise is mainly by the evaporation of sweat (see figure on p. 221). Therefore the rise in core temperature is opposed by a decrease in skin temperature. The average rate of sweat secretion during heavy exercise and peak athletic performance is ca. 1 l/h.

Adjustments to static (postural, holding) work (figure modified from Rohmert, 1962)

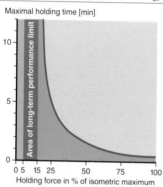

Maximal holding time [min]

Area of long-term performance limit

0 5 15 25 50 75 100
Holding force in % of isometric maximum

Maximal duration of isometric contraction depends very much on contractile force Long-term performance limit is 5–15% of the maximal force (red zone). Maximal force can be exerted for only a few seconds (depletion of anaerobic energy stores, ATP and creatine phosphate, see figure on p. 228).

Muscle blood flow: rises until force reaches ca. 30% of maximal, then it is progressively suppressed by increased intramuscular pressure and is cut off above 70% of maximal force.

Heart rate: increases as static holding work continues. Apparently due primarily to signals in fine muscle afferent fibers. Circulation: especially when holding work involves abdominal compression, venous return is hindered and the central venous pressure rises. Breathing: hyperventilation produced by lactate acidosis and by inputs from muscle receptors.

Performance limits, performance ranges

Long-term performance capacity and limit for dynamic work

Work is within the long-term performance capacity if it can be carried out for at least 8 h without fatiguing the muscles (typical: breathing musculature, heart muscle; light work, see Tables above). Work that produces fatigue has exceeded the long-term performance limit: heavy work (see Tables above) in the range of maximal performance capacity. This kind of performance has a time limit.

Characteristics of fatigue-free long-term performance: constant pulse rate during exercise (below 130/min at age 20–30 years). Recovery pulse sum <100, recovery time <5 min, constant O_2 uptake, O_2 debt <4 liters, blood lactate level <2.2 mmol/l. On bicycle ergometer corresponds to about 100 W for untrained 20- to 30-year-old men, with O_2 uptake of 1.5 l/min.

Athletic performance limit: a useful measure for evaluating performance capacity of top athletes is the blood lactate level. When it rises above 2 mmol/l the long-term performance limit has been reached (see above, aerobic/anaerobic transition); above the anaerobic threshold, at 4 mmol/l even top athletes have reached their performance limit.

Training

Basic terminology of training

Term	Definition/description/comments
Training	Practicing (repeating) similar physical or psychological tasks in order to maintain or increase performance capacity
Training quota	Corresponds to the load in Table on p. 227
State of training	The medium- to long-term adaptation of the organism to the training task as a result of training
Talent	Characteristics that cannot be affected by training but are among the main factors that determine performance capacity (people with little talent achieve only average performance even with intensive training)
Performance limit	Performance plateau finally reached by training at a particular intensity. Performance improves most rapidly at the onset of training, and as training proceeds improvement slows down until the performance limit is reached
Performance maximum	Performance plateau reached by training with the highest possible quota, and which cannot be raised by further training
Forms of training	Only special-purpose training programs give optimal results. Training should be designed both for the particular task (swimming, rowing, etc.) and for the kind of performance (endurance, interval, strength training)
End of training	All adaptations are reversible after training has stopped, and rapidly acquired performance improvements are also rapidly lost. However, learned movement patterns (e.g. swimming, writing) are lost slowly or not at all
Training and age	Trainability, especially of the muscular and cardiovascular systems, decreases progressively with increasing age. Nevertheless, training can prevent, delay or reverse deterioration of performance
Everyday training	Many everyday movements are clearly training stimuli; consider the atrophy caused by muscular inactivity during immobilization (confinement to bed, plaster casts)

Effect of endurance training on physiological parameters in young men
(from Ulmer, 1989b)

Parameter	No training	Endurance training
Heart rate at rest, recumbent (per min)	80	40
Heart rate, maximal (per min)	180	180
Stroke volume at rest (ml)	70	140
Stroke volume, maximal (ml)	100	190
Cardiac output at rest (l/min)	5.6	5.6
Cardiac output, maximal (l/min)	18	35
Heart volume (ml)	700	1400
Heart weight (g)	300	500
Respiratory minute volume, maximal (l/min)	100	200
O_2 uptake, maximal (l/min)	2.8	5.2
Blood volume (liters)	5.6	5.9

Performance tests to evaluate long-term performance capacity

Test	Description/comments
$\dot{V}O_{2max}$	Maximal O_2 uptake. A measure of the aerobic performance capacity of the organism. Measured as ergometer performance increases until exhaustion. Sample values in Table above
PWC_{170}	Pulse working capacity 170. Measurement of the pulse as ergometer performance increases, until pulse rate reaches 170/min. Simple but informative test, suitable for serial studies
Anaerobic threshold	Measurement of blood lactate concentration to determine the aerobic/anaerobic transition and the anaerobic threshold (see above). More informative about endurance in the range of hours than $\dot{V}O_{2max}$. In untrained people the anaerobic threshold is ca. 50–60%, in highly trained endurance athletes about 80% of $\dot{V}O_{2max}$.
Heart volume	Determined by echocardiography or X-rays. A measure of training-related adaptation of the heart to physical endurance activities. Sample values in Table above

High-altitude physiology

Changes in air parameters at various altitudes

Altitude above sea level (m)	Air pressure (mmHg)	Inspired-air P_{O_2} (mmHg)	Alveolar P_{O_2} (mmHg)	Corresponding fraction (O_2/air) at sea level
0	760	149	105	0.2095
2 000	596	115	76	0.164
3 000	526	100	61	0.145
4 000	462	87	50	0.127
5 000	405	75	42	0.112
6 000	354	64	38	0.098
7 000	308	55	35	0.085
8 000	267	46	32	0.074
10 000	199	32		0.055

Functional thresholds for altitude-related O_2 deficiency (after Ruff and Strughold, 1957)

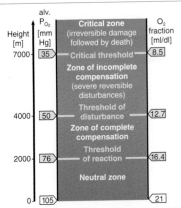

In the neutral zone there is no effect on peak performance capacity. In the zone of complete compensation, heart rate, cardiac output and respiratory minute volume are slightly increased even at rest. During exercise the increase is greater than at sea level. Performance capacity is thus reduced. When the disturbance threshold (safety limit) is passed, performance capacity and the ability to make decisions and react are seriously impaired, and consciousness may be affected. Above the critical threshold, life-threatening CNS disturbances appear, with unconsciousness and seizures; eventually altitude death.

There are broad transitions between the zones. The time course of hypoxia is also important. A rapid loss of pressure in an airplane, causing acute hypoxia, has more dramatic symptoms than slowly developing hypoxia. The symptoms of O_2 deficiency, called altitude sickness, typically include reduced willpower, somnolence, loss of appetite, shortness of breath, tachycardia, vertigo, vomiting, headaches and apathy, but also euphoria (high-altitude intoxication, above 3000 m, resembles alcohol intoxication).

Time from sudden onset of O_2 deficiency to loss of useful function at altitudes of 7000 m or higher (figure modified from Rohmert, 1962)

Altitude (km)	7	8	9	10	11	12	15
Time (min)	5	3	1.5	1	2/3	1/2	1/6

Short-term modifications at high altitudes: adjustments in the range of hours

System	Adjustment/comments
Circulation	See zone of complete compensation in figure p. 233. Arterial blood pressure is not changed during exercise at high altitude. Danger of pulmonary edema (vasoconstriction in lungs due to hypoxia, with increased pressure in pulmonary artery)
Breathing	Only slight increase in respiratory minute volume, because hypoxia is a weak respiratory drive (see p. 201; at 5000 m 10% above sea level value, at 6500 m 100% increase)
O_2 transport	Arterial P_{O_2} decreases with alveolar P_{O_2} (see Table above). At 2000 m, however, O_2 saturation is still 93%. Respiratory alkalosis (see below) shifts O_2 dissociation curve to the left (good for O_2 uptake in the lungs, bad for release in tissues; see p. 205)
Acid–base balance	The hypoxia-induced increase in respiratory minute volume causes respiratory alkalosis (RQ temporarily >1). The base excess (BE, see p. 209) is unchanged

Acclimation to high altitude: long-term adjustments (after Hurtado, 1964)

All values under resting conditions	4540 m above sea level	Sea level
Blood:		
Erythrocytes (million/µl)	6.44	5.11
Reticulocytes (thousand/µl)	46	18
Thrombocytes (thousand/µl)	419	401
Leukocytes (thousand/µl)	7.0	6.7
Hematocrit (%)	60	47
Hemoglobin (g/dl)	20.1	15.6
Blood volume (ml/kg)	101	80
Plasma volume (ml/kg)	39	43
pH, arterial blood	7.39	7.41
Buffer base (mmol/l)	45.6	49.2
Other values:		
Respiratory minute volume (BTPS, l/min per kg)	0.19	0.13
P_{O_2}, alveolar (mmHg)	51	104
P_{CO_2}, alveolar (mmHg)	29.1	38.6
Arterial O_2 saturation (%)	81	98
Heart rate (beats per min)	72	72
Blood pressure (mmHg)	93/63	116/96

The above values have been obtained in serial studies of people living at high altitudes (the Andean city Morococha) in comparison to lowland dwellers (Lima). In centuries of selection, people living in the mountains have acquired the best possible acclimation to high altitudes. But good altitude acclimation can be achieved in only a few weeks, reaching final levels in months or years. The increased Hb content keeps the O_2 transport capacity of the blood nearly constant up to 5000 m altitude. The respiratory alkalosis is compensated by the kidney, by enhanced excretion of bicarbonate. The musculature also adjusts to altitude, by increased capillary density and changes in enzyme levels (to promote aerobic metabolism at low P_{O_2}).

Physiology of diving, life at high pressure

Apneic diving (without breathing apparatus) and snorkeling

Shallow dives
In general, no problems. Excessive prior hyperventilation can cause fainting (blackout) at the end of the dive, because the effect of Pco_2 dominates over the lowering of Po_2, and hence ventilation is not stimulated.

Snorkeling
Possible and permitted only at the water surface. Lengthening the snorkel introduces a pressure difference between the extra- and intrathoracic parts of the low-pressure system, with displacement of blood into the thorax. Danger of lethal overstretching of the blood vessels in thorax and heart.

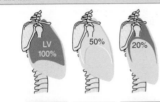

Depth:	0 m	10 m	40 m
Pambient:	1 bar	2 bar	5 bar
LV:	5.0 l	2.5 l	1.0 l
$P_{A_{O_2}}$:	105 mm Hg	210 mm Hg	525 mm Hg

Apneic deep diving:
According to the gas laws (see p. 194), all air-filled body cavities are compressed (lung volume [LV]). Risk of pressure damage: barotrauma, e.g. when pressure in the air-filled middle ear of a person with a cold cannot be relieved by air flow to the nasopharynx due to swelling of the mucosa.

Because alveolar Po_2 decreases during the ascent (from right to left), ascending after staying too long at depth inevitably leads to 'hypoxic' unconsciousness.

Diving with equipment

Method	Description/comments
Compressed air method	In a chamber (caisson) or with portable equipment. Air at the ambient pressure is inhaled; the exhaled air passes into the water (open system). Respiratory minute volume (BTPS, see p. 194) is about equal to that during a similar exercise on land, but, when converted to STPD, becomes greater due to the pressure difference. Breathing is harder because compressed air is more dense. Intoxication (euphoria, etc. as at high altitudes), below 40 m is due to effect of nitrogen on the CNS. The bends result from too-rapid decompression during the ascent, so that gas bubbles form in blood and tissues, causing barotrauma. Recommended stages of decompression are provided in standard tables. In cases of barotrauma, recompression is done in pressure chambers
Oxygen method	Closed system, inspiration of pure O_2, CO_2 absorbed by soda lime. Not suitable for sport diving, because pure O_2 is toxic to the CNS below 7 m. Good for special tasks requiring long-term submersion near the surface
Mixed-gas method	Addition of compressed air or helium to pure O_2. Compressed air allows deeper dives, 7–70 m; helium protects against intoxication

Structure and general characteristics of the GIT

Structure of the GIT

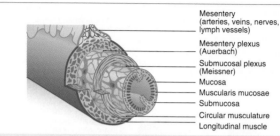

Mesentery
(arteries, veins, nerves,
lymph vessels)

Mesentery plexus
(Auerbach)

Submucosal plexus
(Meissner)

Mucosa

Muscularis mucosae

Submucosa

Circular musculature

Longitudinal muscle

The GIT is a complex hollow structure which stretches continuously from mouth to anus. It is lined by the mucosa (epithelium and corium), in which the epithelial cells are largely responsible for secretion (together with accessory glands) and absorption. The motor component of the GIT consists of two tubes of smooth muscle, one inside the other. In the outer tube the muscle cells are arranged longitudinally, in the inner one they are circular. These two muscle tubes regulate the diameter of the GIT and endow it with motility. Between the two tubes is part of the intestinal nervous system, the myenteric plexus (of Auerbach). A third, very thin muscle tube is located in the submucosa. It serves to move the mucosal villi (pressing the contents to the villi into the efferent lymph and blood vessels). Between the inner and submucosal muscle tubes is the submucosal plexus (of Meissner), another part of the intestinal nervous system.

The smooth musculature of the GIT is either tonically active or phasic–rhythmic (with intermediate forms), depending on its location. The main characteristics of and differences between these two types of muscle are:

	Tonic type	Phasic type
Location, function	In organs with storage function (e.g. gastric fundus, gallbladder, colon) to adjust GIT diameter to the volume to be stored	In most parts of the GIT; basal organ-specific rhythm (BOR) in each place; serves to mix and to transport chyme
Electrical activity	Resting potential −60 to −80 mV. Plateau action potentials like those in myocardium, with Ca influx during the plateau. Spikes can be superimposed on the plateau	Resting potential −60 to −80 mV, usually undulating in slow BOR waves; during depolarization phases, rhythmic Ca spikes with period of seconds are often superimposed
Triggering of contraction	Predominantly by acetylcholine (i.e. under chemical control), also by plateau action potentials plus superimposed spikes (see above)	Predominantly spontaneous activity (myogenic excitation) with no neural or hormonal driving; also Ca spike-triggered contractions
Neural control	Cholinergic nerves stimulate (acetylcholine), adrenergic inhibit (norepinephrine, β-receptors); partly also α-adrenergic activation	Cholinergic nerves stimulate (acetylcholine), adrenergic inhibit (norepinephrine, β-receptors)
Humoral control, pharmacology	Various GI hormones and peptides (see below); resistant to Ca channel blockers (e.g. nifedipine)	Various GI hormones and peptides (see below). Ca channel blockers inhibit Ca spikes, suppress contraction

Elements of the GIT, chyme retention time, secretion rate of digestive glands

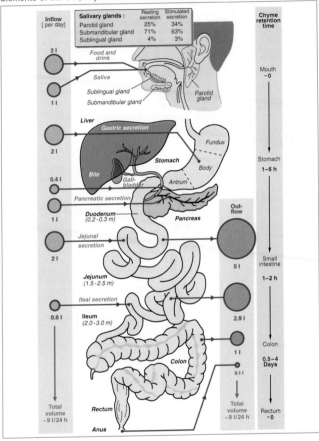

Salivary glands :	Resting secretion	Stimulated secretion
Parotid gland	25%	34%
Submandibular gland	71%	63%
Sublingual gland	4%	3%

Inflow [per day]

2 l — Food and drink

1 l — Saliva / Sublingual gland / Submandibular gland / Parotid gland

Liver — Gastric secretion — Fundus

2 l — Bile — **Stomach** — Body

0.4 l — Gall-bladder — Antrum

Pancreatic secretion

1 l — **Duodenum** (0.2-0.3 m) — **Pancreas**

Jejunal secretion

2 l — **Jejunum** (1.5-2.5 m)

Ileal secretion

0.6 l — **Ileum** (2.0-3.0 m)

Colon

Rectum

Anus

Total volume ~9 l/24 h

Out-flow

5 l

2.9 l

1 l

0.1 l

Total volume ~9 l/24 h

Chyme retention time

Mouth ~0

Stomach 1–6 h

Small intestine 1–2 h

Colon 0.5–4 Days

Rectum ~0

Motility of the GIT

Mastication

Rhythmic reflex by which food is broken into particles of only a few mm³. The chewing movement is elicited reflexly (in some cases also voluntarily) when food is taken into the mouth and touches palate and teeth. A single mastication cycle lasts 0.6–0.8 s, and involves the following forces: incisors 100–200 N, molars 300–900 N (maximum 1500 N). Tongue and cheeks hold the bite between the chewing surfaces. The motor centers for mastication are located in the brainstem.

Swallowing: sequence of three phases (oral, pharyngeal, esophageal) of swallowing; pressure changes during swallowing
(modified from Ewe and Karbach, 1989)

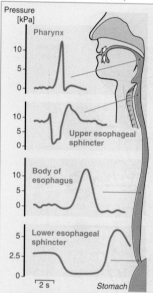

Oral phase: Tip of tongue pushes a portion of the chewed bolus to the base of the tongue and hard palate. Mechanosensors there (glossopharyngeal and vagus nerves) induce an involuntary swallowing reflex.

Pharyngeal phase: The bolus is put under pressure by the muscles in the floor of the oral cavity (efferent innervation: trigeminal, facial, hypoglossal, vagus nerves), the nasopharyngeal space is closed off by the raised soft palate and the larynx by the epiglottis. The bolus slides over the epiglottis into the esophagus, after the upper esophageal sphincter has opened (see recording).

Esophageal phase: The bolus is pushed down by a peristaltic wave (see recording) and after ca. 9 s reaches the lower esophageal sphincter, which relaxes in time to let it pass into the stomach and then closes again (primary peristalsis, see recording).

Liquids usually flow directly

The motor centers for the swallowing reflex are located in the brainstem and in the medulla oblongata. They send out a complicated impulse pattern that controls the act of swallowing. The esophagus is supplied by the vagus nerve.
Secondary peristalsis is produced by mechanical stimulation of the esophageal wall by food fragments (or examination catheter). Transfer of the bolus from esophagus into stomach is facilitated by a brief relaxation of the gastric fundus: receptive relaxation (see also adaptive relaxation on next page).

Gastric motility 1: storage in the fundus (and to some extent in the body)

The solid components of the food stack up in layers, while liquid and gastric juice flow away along the inner wall, to the distal stomach. The pressure inside the stomach increases, but the muscle tone (musculature of the tonic type, see above), continually adjusts to the volume increase reflexly (by a process known as adaptive relaxation; mediated by pressure sensors in the stomach wall, and afferents and efferents in the vagus nerve). The remaining internal pressure and slow waves of contraction slowly propel the soft mass along.

Gastric motility 2: mixing and further transport of the chyme in body and antrum (after Schiller, 1983)

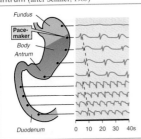

Main function of the stomach: converting the ingested food to a semiliquid mass (chyme) and sending it into the small intestine in portions suited to its processing capacity.

Upper zone of body is the pacemaker region for peristalsis; BOR of the slow waves is 3–4/min. The resulting contraction waves mix and grind the chyme, while the pylorus is closed.

Portions are transferred to the duodenum as soon as the particle size is <2–3 mm (at transfer, 90% is actually (<0.25 mm).

Gastric motility 3: factors determining the rate of gastric emptying.
Food remains in the stomach between 1 and 6 h, depending on:

I. Composition of the food

1   **Consistency:** Solid and coarse food stays longer than liquid (comminution by stomach peristalsis takes time)

2   **Osmolarity:** Isosmotic fluids empty most rapidly. The higher or lower the osmolarity, the longer the food is retained

3   **Nature and energy content of the food:** Time in stomach increases in the order carbohydrates > proteins > fats

II. Composition of the chyme

Is measured by chemosensors in the duodenum, in order to adjust the stomach-emptying peristalsis to the 'processing capacity' of the small intestine, by neural and humoral signals.

1   **Acidity:** Acidic chyme slows down gastric emptying (pH optimum for the enzymes in small intestine is 7–8; it is adjusted by pancreatic secretion and secretion of the Brunner's glands in the small intestine)

2   **Osmolarity:** Higher osmolarity of the chyme slows down gastric emptying (in the duodenum the chyme is made isotonic, and it remains so during passage through the small intestine; when osmolarity is high, this process takes longer)

3   **Fat content:** High content of fat in the chyme slows down gastric emptying

Motility of the small intestine

1 Non-propulsive peristalsis	Local, circular waves of contraction to mix the chyme
2 Rhythmic segmentation	Local constrictions separated by 10–20 cm, to mix the chyme; a slight propulsive component is also present
3 Pendular movements	Of the longitudinal musculature, to mix the chyme
4 Propulsive peristalsis	Propels the chyme toward the large intestine; basic rhythm of the associated pacemaker activity (slow waves, BOR) in duodenum 12/min (see figure on preceding page); frequency decreases at the ileum end, to 8/min
5 Interdigestive activity complexes	Particularly strong propulsive peristalsis between meals, at intervals of over an hour; may serve as a mechanism for general cleaning of the GIT

Neural control of motility of the small intestine (figure based on Mayer, 1974)

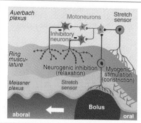

Local, descending, inhibitory peristaltic reflex. Stimulation of stretch sensors in the submucosal plexus activates inhibitory motoneurons in the myenteric plexus, which inhibit the aboral circular musculature. This relaxation enables propulsion of the bolus by contraction of the circular muscles proximal to the bolus.

The neuromuscular transmitters of the inhibitory neurons are apparently NO (nitric oxide), ATP and VIP.

Local, ascending, excitatory peristaltic reflex; actions opposite to those of the inhibitory reflex; enhances myogenic contractions. The neuromuscular transmitters of the excitatory motoneurons are acetylcholine (muscarinic receptors) and substance P (perhaps also other peptides, see also the Table on the next page). The parasympathetic system has a generally facilitatory action on intestinal motility, the sympathetic system is inhibitory.

Motility patterns of the large intestine

1 Non-propulsive motility	Circular contractions occurring simultaneously at several places, to mix the chyme
2 Peristaltic waves	Rare, advancing contractions preceded by relaxation, which propel the chyme about 20 cm forward
3 Mass movements	Occur only three to four times per day; propel the chyme for long distances

The neural control of movements of the large intestine is like that in the small intestine (see above). Disturbances of colonic motility cause constipation and diarrhea, e.g. as in irritable bowel syndrome (IBS).

Peptides and other neurotransmitters with known function in the GI tract
(after Furness *et al.*, 1992)

Substance	Localization/function/comments
ACh	Acetylcholine; most important excitatory transmitter to muscle, to intestinal epithelium, to gland cells, to some endocrine cells of the GIT and at neuroneuronal synapses
ATP	Adenosine triphosphate; co-transmitter in inhibitory motoneurons (see preceding page)
CCK	Cholecystokinin; in some GIT neurons; may be involved in excitatory transmission; in general excites smooth muscle
GRP	Gastrin-releasing peptide; corresponds to the bombesin of amphibia; excitatory transmitter for gastrin cells
NPY	Neuropeptide Y; inhibitory transmitter at gland cells
NO	Nitric oxide; cotransmitter of inhibitory motoneurons (see preceding page); perhaps also at neuroneuronal synapses
NA	Norepinephrine; NAergic nerve fibers in GIT are all sympathetic; inhibit motility in non-sphincter regions, contract sphincter muscles, inhibit secretomotor reflexes, have vasoconstrictor action on intestinal vessels
Serotonin	5-HT; transmitter at some excitatory neuroneuronal synapses
Tachykinins	These include: substance P (SP); neurokinin A (NKA); neuropeptide K; neuropeptide Y; excitatory transmitters to smooth muscle and at some neuroneuronal synapses; co-transmitters with ACh
VIP	Vasoactive intestinal peptide; excitatory transmitter at gland cells; may be transmitter of vasodilator neurons; co-transmitter at inhibitory motoneurons particularly in the lower esophageal sphincter (see preceding page)

Secretion, digestion, absorption in the GIT

Hormones and candidate hormones in the GIT (see also above Table)

Substance	Presence/action/comments
Gastrin	Is released from G-cells in the antrum (pyloric region) and duodenum in response to the food mass; triggers HCl secretion (together with ACh and histamine, see p. 243); promotes antrum motility; trophic action on gastric mucosa
CCK	Released from duodenum and jejunum; elicits gallbladder contraction and pancreatic secretion (enzymes); trophic action on pancreas
Secretin	Released from duodenum and jejunum; elicits pancreatic secretion (bicarbonate-rich); inhibits secretion of gastric acid
GIP	Gastric inhibitory polypeptide; released from duodenum and jejunum; increases insulin release; inhibits secretion of gastric acid and gastrin
Somatostatin	From entire GIT; inhibits secretion in stomach and pancreas
Enteroglucagon	From ileum and colon; inhibits secretion in stomach and pancreas; trophic action on intestine
Motilin	From duodenum and jejunum; stimulates muscle contraction
Neurotensin	From ileum; inhibits secretion of gastric acid

Secretion of saliva (see also Fig. p. 237) (after Thaysen *et al.*, 1954)

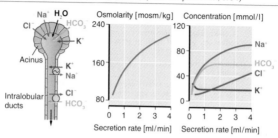

The first step in saliva production is the transcellular transport of Cl^- into the acinar lumen; Na^+ ions follow because of the lumen-negative voltage and H_2O for osmotic reasons. The primary saliva formed by the acini has an electrolyte composition similar to that of plasma; in passage through the ducts leaving the acini it is secondarily modified, the more so, the slower the saliva flows (read curves from *right* to *left*).

Functions of saliva: (1) makes the food bolus smooth and swallowable by adding mucus (mucopolysaccharides, mainly from submandibular and sublingual glands); (2) enhances the taste sensation by dissolving and suspending the food (which reflexly induces additional salivary secretion and secretion by GIT glands); (3) initiates digestion of starches (to disaccharides: maltose, glucose) by α-amylase (mainly from the parotid glands); (4) keeps the mouth moist between meals and facilitates speaking; has a cleaning and antibacterial action and protects the teeth from caries.

Neural control of salivary secretion: Elicited mainly by reflexes, both unconditioned (strongest stimulus: acidic liquids) and conditioned (Pavlovian); control centers in medulla oblongata; efferent elements include parasympathetic (ACh, substance P, stimulates abundant low-viscosity saliva from parotid gland) and sympathetic (α-adrenergic, stimulates small amounts of viscous saliva from the other glands).

Structure of the tubular glands in the gastric mucosa; components of gastric juice

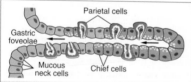

Thickness of the folded gastric mucosa is 0.6–0.9 mm (bounded by columnar epithelium); its glands (general structure in figure) secrete 2–3 liters of gastric juice daily. The superficial mucous cells produce mucus (which is deposited in a layer 0.6 mm thick) and bicarbonate, the lower-lying parietal cells, hydrochloric acid (HCl) and intrinsic factor and the chief cells, pepsinogen. Gastrin and other hormones are released from endocrine cells see Table on p. 241.

Phases of induction of gastric juice secretion

Phase	Description/comments
Fasting	Slight secretion of a viscous gastric mucus; stops after vagotomy and removal of the antrum (site of the G-cells), i.e. basic vagal tone is responsible for this basal secretion
Cephalic	Secretion induced by CNS, due to both mental (expectation, imagination) and sensory stimuli (sight, smell, taste); efferent mediation by vagus nerve (ACh, muscarinergic); disappears after denervation of the antrum, hence it is initiated indirectly by ACh-induced gastrin release; up to 55% of the maximal possible secretion can be so evoked
Gastric	Secretion reflexly triggered by the food mass: (1) mechanosensors activate (a) short intramural reflex paths, (b) reflexes with afferents and efferents in vagus nerve; (2) chemical stimuli (peptides, amino acids, alcohol, caffeine) induce release of gastrin, which stimulates secretion hormonally
Intestinal	Entry of chyme into small intestine, depending on its composition, causes mechanical and chemical (amino acids) stimuli that trigger the release of secretin and other hormones

Phases of inhibition of gastric juice secretion

Phase	Description/comments
Cephalic	Stress can both excite and inhibit secretion; inhibition occurs by way of the sympathetic system, mechanism unknown
Gastric	Gastric contents with pH <3 inhibit gastrin release and thus suppress acid and pepsinogen production; for influence on gastric motility, see p. 241
Intestinal	Entry of chyme into small intestine, depending on its composition, causes the release of various hormones (see p. 241) that inhibit the release of gastric acid and gastrin. Fats have the strongest inhibitory action

Induction and cellular mechanism of gastric acid secretion
(from the results of many authors, based on Schmidt and Thews, 1989)

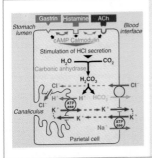

The secretion of hydrochloric acid by the parietal cells is controlled by three receptor systems. There is a strong, obligate interaction between the histamine (H_2) and gastrin receptors and a weaker, facultative one between H_2 and ACh receptors (arrow thickness). Hence, blockade of the H_2 receptor decreases gastrin- and ACh-stimulated secretion.

Second messengers are cAMP and Ca^{2+}/calmodulin.

Secretion process uses ATP-driven active transport of H^+ via an electroneutral H^+/K^+ exchange pump (H^+,K^+-ATPase) in parallel with Cl^- channels; K^+ recycles via K^+ channels (net effect, HCl secretion).

Maximal acidity ca. pH 1, usually 1.8–4.

Pancreas: topography, relation to bile ducts, amounts of exocrine secretion

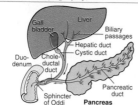

Weight of the pancreas ca. 110 g; can produce 1–1.5 l/d of secretion, i.e. about 10 times its own weight.

The main outlet, the pancreatic duct or duct of Wirsung, passes through the whole organ and opens into the duodenum next to or (in 30–40% of people) together with the common bile duct through the sphincter of Oddi (the major duodenal papilla).

For endocrine pancreas see p. 144.

Composition of the pancreatic juice 1: electrolytes and water (after Bro-Rasmussen *et al.*, 1956, and Ewe and Karbach, 1989)

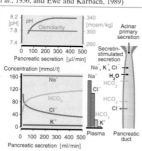

The pancreatic juice is alkaline because of its high bicarbonate content; together with bicarbonate from Brunner's glands, alkaline mucus from goblet cells and bile it serves to neutralize the chyme, see p. 247.

The secretion is isosmotic, regardless of the rate of secretion.

Water and bicarbonate come from the ductal epithelial cells. Fluid secretion occurs principally in the ducts and involves intracellular hydration of CO_2 ($CO_2 + H_2O$ [carbonic anhydrase] → H^+ + HCO_3^-). The H^+ is exchanged for Na^+ at the basolateral cell membrane and the HCO_3^- is secreted (probably in exchange for Cl^-). Na^+ and K^+ are secreted via the intercellular junctions because of the lumen-negative voltage.

Composition of the pancreatic juice 2: digestive enzymes

Type of enzyme	Description/comments
Peptidases	For protein digestion; main representatives (among many others) are the proenzymes trypsinogen and chymotrypsinogen; enterokinases from the duodenal mucosa activate trypsinogen to trypsin, which activates chymotrypsinogen to chymotrypsin
Amylases	For carbohydrate digestion; main representative α-amylase, is secreted in active form
Lipases	For fat digestion; main representatives are pancreatic lipase and phospholipase A. Colipase enhances lipase action by providing a binding site on fat droplets for pancreatic lipase
Nucleases	Main representative ribonuclease; splits nucleotides off from ribonucleic acids

The enzymes come from the acinar cells; their granules contain the enzymes in constant proportions, the largest fraction consisting of peptidases. 90% of the proteins in the pancreatic juice are enzymes. Secretion of the peptidases as proenzymes protects the pancreas from self-digestion. Stimulation of the acinar cells generates a small amount of concentrated secretion; stimulation of epithelial cells produces larger volumes of watery secretion (see above).

Phases of induction of pancreatic juice secretion

Phase	Description/comments
Fasting	In the basal secretion at rest, bicarbonate production is only 2–3% of maximum possible; enzyme secretion is 10–15% of maximal
Cephalic	Elicited by stimuli like those in the cephalic phase of gastric juice secretion, see above; secretion rates increase by 10–15% for bicarbonate (transmitter: VIP), 25% for enzymes (transmitter: ACh)
Gastric	Stretching of corpus and antrum of the stomach evokes vagovagal gastro-pancreatic reflexes; increased bicarbonate and enzyme secretion with only moderate increase in flow
Intestinal	Entry of acidic chyme into the small intestine causes release of CCK from I cells (action on enzyme secretion) and secretin from S cells (action on bicarbonate secretion); these are the two strongest stimulators of pancreatic secretion, see also p. 241

Phases of inhibition of pancreatic juice secretion

Phase	Description/comments
Cephalic	By way of sympathetic system, like cephalic inhibition of gastric juice secretion
Gastric	All factors that reduce the rate of stomach emptying (see p. 239) also inhibit the secretion of pancreatic juice, via the changes in release of CCK and secretin
Intestinal	A high fat content of the duodenal chyme inhibits pancreatic secretion; the mechanism is unknown

Liver: secretory function of the hepatocytes and the bile duct epithelium

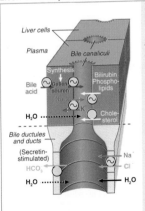

Hepatocytes, the parenchymal cells of the liver, actively secrete a number of organic molecules into bile. The most important of these are: (1) bile acids, (2) cholesterol, (3) phospholipids, (4) bilirubin conjugates. The hepatocytes are epithelial cells which separate the blood space from the biliary space. The sinusoidal membrane domain faces the blood and the canalicular membrane domain lines the bile canaliculi (terminal branches of the biliary tree). These two membrane domains are separated by tight junctions, as in all epithelia. Bile acids and bilirubin are taken up across the sinusoidal membrane by specific transporters and secreted across the canalicular membrane by ATP-dependent pumps. Electrolytes (Na^+, Cl^-, HCO_3^-) equilibrate across either tight junctions (Na^+) or the canalicular membrane (Cl^-, HCO_3^-) but are not the driving force for hepatic bile formation. Water equilibrates passively across cell membranes and tight junctions.

The bile duct epithelium comes in contact with the bile after it exits the canaliculi. This is a site of HCO_3^-/Cl^- exchange. The bile duct epithelium alkalinizes the bile and contributes to total bile secretion.

Composition of hepatic and gallbladder bile (Table after Ewe and Karbach, 1989)

Constituent	Hepatic bile (mmol/l)	Gallbladder bile (mmol/l)
Na⁺	165	28
K⁺	5	10
Ca²⁺	2.5	12
Cl⁻	90	15
HCO₃⁻	45	8
Bile acids	35	310
Lecithin	1	8
Bile pigments	0.8	3.2
Cholesterol	3	25
Bilirubin	1.5	15
pH	8.2	6.5

The hepatic bile is colored golden yellow by bilirubin; each day 600 ml is produced, 50% of which flows directly into the small intestine; the rest flows to the gallbladder, Figs pp. 237, 244.

However, the capacity of the gallbladder is only 50–60 ml. More bile can be accepted because of the rapid absorption of salt and water by the gallbladder epithelium (primary process: active transport of NaCl out of the bile). Due to absorption of HCO_3^- the pH falls to 6.5 (see Table).

The bile acid molecules have hydrophilic and hydrophobic domains and can therefore act as detergents for the digestion of fats (see figure on p. 248). Hence they facilitate the emulsification of the fats in the chyme and by forming micelles, molecular aggregates incorporating fatty acids, make it possible for the fatty acids to be subsequently absorbed in the intestinal cells (for circulation of bile acids see figure below). Bilirubin is the product of hemoglobin breakdown, i.e. the means by which Hb is excreted from the body (200–300 mg/d). Intestinal bacteria in the terminal ileum and in the colon convert it into the darker urobilinogen, which is responsible for the color of the stool.

Regulation of bile secretion and delivery to the intestine (see also p. 241)

Substance	Action/comments
CCK	Cholecystokinin causes (1) contraction of the gallbladder and (2) relaxation of the sphincter of Oddi; all processes leading to release of CCK therefore increase the flow of bile into the duodenum
Vagus nerve	Action similar to but weaker than that of CCK; the same applies to cholinergic drugs
Secretin	Stimulates secretion in the bile ductules and ducts, see figure on p. 245 (action identical to that on epithelial cells of pancreas)
Bile acids	Secretion of bile acids by hepatocytes increases when the level of bile acids in the blood plasma increases, and vice versa; all processes that increase flow of bile into the duodenum therefore indirectly lead to increased bile secretion (see also below)

Enterohepatic circulation of bile acids

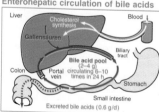

About 90% of the bile acids delivered to the intestine are reabsorbed: half passively (in small intestine and colon) and half actively (only terminal ileum).

The amounts lost in the feces (ca. 10% of the pool of ca. 3 g) are replaced by synthesis in the hepatocytes (for this as well as their conjugation and the formation of bile salts see biochemistry textbooks).

Exocrine and endocrine secretory cells in mucosa of small intestine

Source	Secretion/description/comments
Brunner's glands	Secretion comparable to that of the pancreatic acinar cells (see figure on p. 244), i.e. high concentration of amylases and peptidases with small content of mucin, low Cl^- and high HCO_3^- concentration; isosmotic solution, pH 8.2–9.3; stimuli similar to those for the pancreatic acinar cells, especially those of the intestinal phase
Goblet cells	Found throughout large and small intestine; produce alkaline mucus consisting of various glycoproteins; this acts together with the Brunner's gland secretion, the bile and the pancreatic juice to neutralize the acidic gastric chyme
Endocrine cells	The classical GI hormones (e.g. gastrin, secretin, CCK) are listed on p. 241. These are all (also) released from the mucosa of the small intestine

Absorption of foodstuffs along the small intestine: a survey

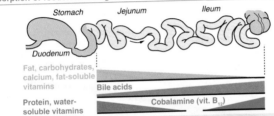

NaCl and water absorption in the small intestine (for water balance, see p. 237) occurs in the villus cells and involves Na^+/H^+ exchange and Cl^-/HCO_3^- exchange at the luminal cell membrane and Na^+ pumping (Na^+,K^+-ATPase) and Cl^- extrusion via channels at the basolateral membrane. Water follows because of the changes in osmolality produced by NaCl transport. In the stomach only certain ions (e.g. Na^+) and lipid-soluble substances, mainly alcohol, are absorbed. In the colon there is net absorption of the NaCl and secretion of K^+. In contrast with the small intestine, Na^+ entry into the epithelial cells is via luminal membrane Na^+ channels; K^+ secretion is also via channels in this membrane. Na^+ absorption and K^+ secretion in colon epithelium are sensitive to mineralocorticoids.

Structure of the villi, where all absorption in the small intestine occurs

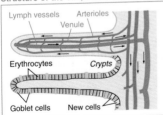

The enterocytes (life-span 3–6d, replaced by cells originating in the crypts) are the absorbing apparatus of the small intestine (surface area ca. $200\,m^2$).

The vascular system in the villi assists in the uptake of water, electrolytes and foodstuffs from the intestinal lumen, by the counter-current principle (p. 256).

Digestion and absorption of carbohydrates in the small intestine
(after Birbaumer and Schmidt, 1996)

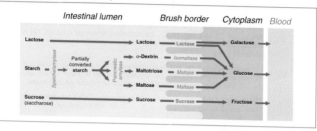

Digestion and absorption of fats (after Birbaumer and Schmidt, 1996)

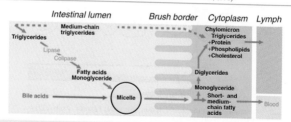

The fat components of the micelles dissolve in the lipid membrane of the enterocytes and diffuse into the cell. The micelle then takes up new fat components. In the enterocyte intracellular lipid synthesis takes place, after which the fat is incorporated into lipoproteins (formation of chylomicrons) and exocytosed into the central lymph duct of the intestinal villus.

Digestion and absorption of proteins (after Birbaumer and Schmidt, 1996)

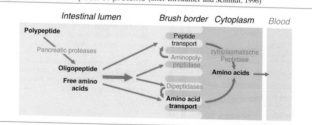

Neural control of intestinal evacuation

Intestinal continence and defecation (after Birbaumer and Schmidt, 1996)

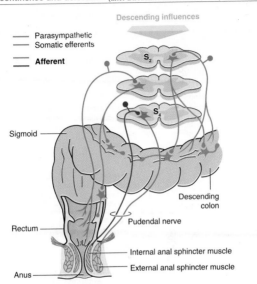

Intestinal continence and defecation are the most important functions of the rectum and the anus. The figure gives an overview of the structures and neuronal circuits involved.

Continence: The rectum is closed externally by two circular muscles (internal and external anal sphincter muscles). The internal is a smooth muscle (no voluntary control is possible), and the external is striated (hence can be controlled voluntarily).

Filling of the rectum with intestinal contents reflexly induces relaxation of the internal sphincter muscle along with increased contraction of the external sphincter. At the same time the urge to defecate is felt. If defecation does not occur, the rectum adjusts to its increased contents and the relaxation of the internal sphincter decreases (filling capacity of the rectum ca. 2 liters).

Defecation: Reflex contraction of the end sections of the colon with simultaneous relaxation of both sphincter muscles, normally with voluntary assistance (abdominal musculature tensed, diaphragm lowered by contraction of the thoracic musculature with lungs inflated and glottis closed).

Survey of functions, structure, mechanisms

The kidneys have two main functions

1. To excrete most of the non-volatile waste products of the body's metabolism (excretory function).
2. To keep constant the volume and the electrolyte composition of the extracellular fluid (homeostatic function).

They thus provide every cell in the body with a constant environment (internal milieu).

Anatomy of the kidney (adapted from Birbaumer and Schmidt, 1996)

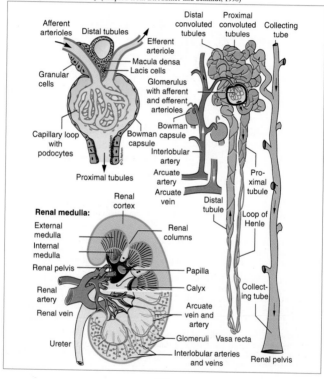

The functional units in the kidney are the nephrons; each kidney has ca. 1.2 million nephrons, at least three times as many as are absolutely required

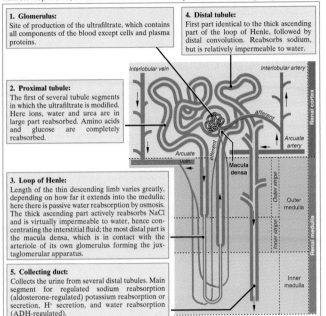

1. Glomerulus:
Site of production of the ultrafiltrate, which contains all components of the blood except cells and plasma proteins.

4. Distal tubule:
First part identical to the thick ascending part of the loop of Henle, followed by distal convolution. Reabsorbs sodium, but is relatively impermeable to water.

2. Proximal tubule:
The first of several tubule segments in which the ultrafiltrate is modified. Here ions, water and urea are in large part reabsorbed. Amino acids and glucose are completely reabsorbed.

3. Loop of Henle:
Length of the thin descending limb varies greatly, depending on how far it extends into the medulla; here there is passive water reabsorption by osmosis. The thick ascending part actively reabsorbs NaCl and is virtually impermeable to water, thus concentrating the interstitial fluid; the most distal part is the macula densa, which is in contact with the arteriole of its own glomerulus forming the juxtaglomerular apparatus.

5. Collecting duct:
Collects the urine from several distal tubules. Main segment for regulated sodium reabsorption (aldosterone-regulated) potassium reabsorption or secretion, H⁺ secretion, and water reabsorption (ADH-regulated).

Pressure changes in the renal vascular system (after Thurau and Wober, 1963)

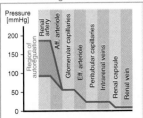

An afferent arteriole enters each glomerulus and branches to form a capillary tuft. The capillaries coalesce into an efferent arteriole, which splits into a second capillary network that perfuses the tubule (see above).

The afferent arteriole is the site of autoregulation of renal blood flow; when the blood pressure rises, it responds by increasing resistance (constriction) just enough to keep the blood flow constant over a broad pressure range (see Fig. p. 252, total renal blood flow p. 176).

Glomerular filtration

Fine structure of the filtration barrier in the glomerulus, hydraulic conductance

The ultrafiltrate emerging from the glomerular capillaries, i.e. the primary urine, must pass the following three barriers (from capillary lumen to urinary space):
1. Endothelial pores in the capillary wall (diameter 50–100 nm).
2. Basement membrane.
3. Slit pores between foot processes of podocytes (epithelium of Bowman's capsule) (diameter 20–50 nm).
The permeability to water, i.e. the hydraulic conductance of the glomerular capillaries, is considerably higher than that of most other capillaries in the body. Furthermore, glomerular capillaries are permeable to molecules up to about 30000 daltons, though permeation is progressively impeded from 10000 daltons on. Plasma proteins of MW 60000 or more and cells are impermeable. (For ultrastructure see textbooks of histology.)

Pressures that determine the glomerular filtration rate (GFR)

These are the same pressures that determine filtration in other capillaries (see p. 184). When applied to the glomerular capillary, the equation for effective filtration pressure p_{eff} is

$$p_{eff} = p_c - p_u - \Pi_c \qquad (1)$$

where p_c is the capillary hydrostatic pressure (ca. 50 mmHg), p_u is the pressure in the urinary space, i.e. Bowman's capsule (equivalent of interstitial pressure, ca. 12 mmHg) and π_c is the colloid osmotic (oncotic) pressure of the blood (ca. 20 mmHg) at the beginning of the capillary; on the tubule side there is no significant oncotic pressure, because virtually no proteins pass through the filter; hence

$$p_{eff} = 50 - 12 - 20 = 18 \text{ mmHg at the beginning of the capillary}$$

Along the glomerular capillaries p_c decreases slightly (to ca. 48 mmHg), but π_c becomes much greater due to the increase in protein concentration as ultrafiltration takes place. If $\pi_c = p_c - p_u$, i.e. $\pi_c \cong 38$ mmHg, then filtration would stop ($p_{eff} = 0$, filtration equilibrium). In man, p_{eff} does not fall to 0 under normal conditions.

Dependence of GFR on renal blood flow (RBF)

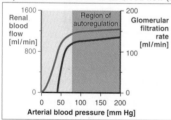

GFR depends strongly on the RBF. The mechanism is the relationship between glomerular blood flow and the rate of rise of π_c (at high flow rates, Π_c rises more slowly), and hence P_{eff} falls less with capillary distance.

Determinants of GFR, filtration fraction GFR/RPF (RPF = renal plasma flow)

The GFR is directly proportional to (1) the P_{eff}, (2) the hydraulic conductance and (3) the effective filtration area, which is a function of the number of glomeruli. The net result is a normal GFR of 120 ml/min, or about 180 l/d. About 1 liter of blood flows through the two kidneys per minute, or about 600 ml plasma (RPF). Of this 600 ml plasma, 120 ml is filtered out, so that the filtration fraction GFR/RPF = 20%.

Measurement of GFR, clearance

Measurement of GFR with inulin or creatinine

An indicator substance suitable for measuring GFR must meet the following four criteria:
1. Freely filterable in the glomeruli.
2. Not reabsorbed or secreted in the tubules.
3. Not synthesized, metabolized or accumulated by the kidney.
4. Not able to influence renal functions.

The amount of such a substance that filters in the glomeruli must appear in exactly the same amount in the excreted urine. Because

$$\text{Amount} = \text{volume} \times \text{concentration} \qquad (2)$$

the urine flow rate (volume of urine per unit time) $\dot{V}$, the concentrations of the indicator in the urine (U) and in the plasma (P), and the GFR are related as follows:

$$U \dot{V} = P \times GFR \qquad (3)$$

$$GFR = U/P \cdot \dot{V} \ [ml/min] \qquad (4)$$

The fructose polymer inulin meets the above criteria and hence can be infused intravenously to determine GFR. An almost equally good indicator is creatinine, an endogenous substance which has a practically constant P (9 mg/l).

Definition of the term 'clearance'

The GFR found with the above method (120 ml/min) is the volume of plasma that has been freed ('cleared') of the indicator substance per min. In this (special) case, GFR equals the clearance of inulin (C_{in})

$$C_{in} = U_{in}/P_{in} \cdot \dot{V} \ [ml/min] \qquad (5)$$

In the case of a substance x that is not only filtered but also secreted in the tubules (e.g. PAH, see below), $C_x > C_{in}$. For substances (y) that are reabsorbed (e.g. Na$^+$), the reverse is true ($C_y < C_{in}$).

In every case, the clearance C represents the (calculated) volume of plasma cleared of a particular substance per unit time.

Clearance of *para*-aminohippuric acid (PAH) as a measure of RPF

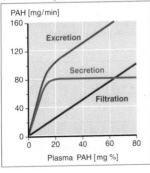

PAH [mg/min]

Excretion

Secretion

Filtration

Plasma PAH [mg %]

PAH is freely filtered and also secreted practically completely in the proximal tubules as long as P remains below saturation levels for the secretory process. Under these conditions, all the plasma flowing through the kidneys is cleared of PAH, so that the C_{PAH} is a direct measure of RPF, and thus is normally 600 ml/min.

Because PAH secretion is an active process, it is saturable; at this P and beyond, PAH secretion remains constant.

Clearance and tubule transport of organic substances

Survey of the localization of transport processes in the nephron

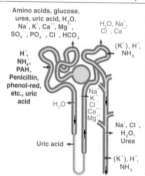

For any excreted substance, C_x/C_{in} = fractional excretion. $C_x/C_{in} = 1$ if the substance is only filtered, >1 if it is also secreted (extreme case: PAH see p. 253), and <1 if it is reabsorbed (extreme case: $C_x = 0$, i.e. complete reabsorption, e.g. of glucose at normal plasma concentration, see figure below).

As the figure shows, the most important site of reabsorption and secretion of organic substances is the proximal tubule.

Clearance of glucose as example of reabsorbed organic substances

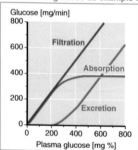

Glucose is filtered at a rate of about 180 g/d, but normally none of it appears in the urine ($C_{glu} = 0$, see p. 255). It is very efficiently reabsorbed by the proximal tubule cells. The mechanism is co-transport with Na^+ across the luminal membrane and carrier-mediated transport of glucose alone from cell to capillary, across the basolateral membrane.

For normal GFR, at plasma concentrations exceeding 250 mg/dl, glucose appears in the urine (glucosuria), and at 400–500 mg/dl the maximal transport capacity is reached.

Reabsorption of amino acids occurs by similar mechanisms and is practically complete, also exclusively in the proximal tubule.

Secretion and excretion

Secretion is the net active transport of substances (e.g. PAH p. 253) from the blood into the lumen. Mainly organic acids and bases are secreted in the proximal tubule (see figure above; for the special case of potassium ions see p. 258). Excretion in the urine is the net result of filtration, reabsorption and secretion.

Handling of oligopeptides and proteins by reabsorption by the proximal tubule

Small amounts of both kinds of substances appear in the filtrate. Oligopeptides are hydrolyzed by peptidases of the brush border membrane, forming amino acids which are practically completely reabsorbed (see above). Proteins are taken into the cells of the tubule by endocytosis, and there they are digested in lysosomes.

Reabsorption of salt and water

Typical amounts filtered and reabsorbed per day (after Patton *et al.*, 1989)

Substance	Unit	filtr. (Load)	Excr.	Reabs. (%)
Water	liter	180	1.5	>99
Na⁺	mmol	25 000	150	>99
K⁺	mmol	630	95	85
Cl⁻	mmol	18 000	150	>99
HCO₃⁻	mmol	4 500	0	100
Glucose	g	180	0	100
Amino acids	g	70	0	100
Urea	g	58	23	60

Na⁺ absorption is coupled directly or indirectly to transport of several other solutes and of water. Na⁺ is reabsorbed in all tubule segments except the thin descending segment of the loop of Henle. The reabsorption of Na⁺ is the most important function of the kidney; 90% of its O_2 consumption is required for this purpose.

Fractional and hormonally controlled Na⁺ reabsorption in the various tubule segments

Tubule section	Process/comments
Proximal	65% of the filtered load is reabsorbed here, regardless of the GFR and without saturation (fractional reabsorption). Na⁺ enters the tubule cell down the favorable electrochemical gradient and is actively transported (Na pump) through the basolateral membrane into the interstitial space. Cl⁻ and HCO₃⁻ reabsorption are carrier-mediated and coupled to Na⁺ reabsorption directly or indirectly. Water is reabsorbed by osmosis, because of the changes in osmolality produced by reabsorption of Na salts. Water reabsorption is ca. 65% of GFR.
Thick ascending limb of the loop of Henle	(Also called pars recta of distal tubule, see pp. 250, 256.) Active reabsorption of Na⁺ continues (fractional reabsorption about 25%). Hence, at the end about 10% of the load remains. Because the tubule wall is impermeable to water, the tubule fluid becomes hyposmotic to plasma: prerequisite for diluted final urine in case of water excess.
Distal tubule and collecting duct	Continued active Na⁺ reabsorption, but now for the first time under the control of hormones: (1) aldosterone, which increases Na⁺ reabsorption and K⁺ secretion (see also p. 258) mostly in distal tubule and cortical collecting duct, and (2) atriopeptin (ANF), which inhibits Na⁺ reabsorption in medullary collecting duct. Na⁺ excretion can vary from ca. 5% to < 0.5% of the filtered load. Regulation of urine volume and osmolality is provided by antidiuretic hormone, which increases permeability of the collecting duct to water, increases water reabsorption, and causes production of a low-volume, high-osmolality urine (see pp. 138, 187).

Renal O_2 consumption is directly proportional to active Na⁺ transport (after Deetjen and Kramer, 1961)

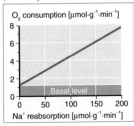

Active Na⁺ reabsorption from the tubule fluid is the key process in renal tubule function. The O_2 consumption of the kidney is directly proportional to the rate of Na⁺ reabsorption.

Mechanisms of concentration and dilution of the urine

Water reabsorption along the nephron (modified from Deetjen, 1989a)

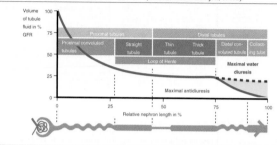

The diagram summarizes the discussion of water reabsorption on the preceding page, namely the large fractional reabsorption in the proximal convolution and thin descending limb of the loop of Henle, the lack of reabsorption in the ascending segments of the loop of Henle (despite large NaCl reabsorption, see above) and the hormone-controlled concentration or dilution of the urine (antidiuresis or water diuresis) in the distal tubules and collecting ducts.

Mechanisms for antidiuresis: (1) osmolality increased in interstitial space of renal medulla by countercurrent multiplication and countercurrent exchange systems and (2) water permeability of collecting duct controlled by ADH levels

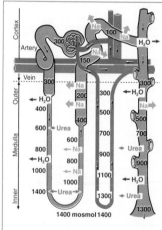

The driving force for the concentrating mechanism is provided by the active transport of NaCl out of the thick ascending segment of the loop of Henle (straight distal tubule) without simultaneous reabsorption of water (see above). This makes the renal medulla hyperosmotic, so that it can drive osmotic water reabsorption in the descending limb of Henle's loop (most of which is carried away by the vasa recta). In the thin ascending limb of Henle's loop there is also net NaCl reabsorption, by a passive process dependent on different permeabilities to urea and NaCl.

The whole process is amplified step by step on the way into the renal medulla by countercurrent multiplication, so that in man the inner medulla becomes four times hypertonic to the blood plasma.

The ascending and descending limbs of the vasa recta constitute the counter current exchange system.

The steep osmotic gradient is used to absorb water from the collecting duct, with ADH increasing water permeability depending on plasma osmolality.

Mechanism of water diuresis, pathophysiology of diabetes insipidus

Water diuresis: Any excess of water is eliminated in normal people within a few hours. The mechanism involves a marked reduction of ADH release because of low plasma osmolality, as a result of which the urine becomes hyposmotic relative to plasma (for regulation of ADH secretion see pp. 138, 189). Alcohol also inhibits the release of ADH, which explains its diuretic action.

Diabetes insipidus: When ADH production is diminished or absent (central diabetes insipidus) or when renal tubule receptors are absent or defective (renal diabetes insipidus), large amounts of hyposmotic urine are excreted (polyuria, in the extreme 20–30 l/d). The resulting thirst causes constant drinking (polydipsia).

Contribution of medullary blood flow to antidiuresis
(modified from Deetjen, 1989a)

	% renal weight	% perfusion	Blood volume ml/g tissue	Perfusion [ml•g^{-1}•min^{-1}]
Cortex	70	92	0.2	5.3
Outer medulla	20	7	0.2	1.4
Papilla	10	1	0.2	0.4

The renal medulla, including the papilla, receives only a small fraction of renal blood flow (cf. p. 176). Furthermore, it contains only loop-shaped capillary vessels (arterial and venous vasa recta) up to several cm in length. They form a vascular countercurrent system (countercurrent exchange, see figures on pp. 250, 256) that functionally supplements the countercurrent multiplication system of the tubules by limiting washout of solute while removing water from the interstitial space.

Special forms of tubule transport

Transport of waste products of protein metabolism

Urea: 25–30 g/d, hence the main waste product of protein metabolism (non-toxic, inert, electrically neutral, plasma level 15–40 mg/dl or 2.5–6.7 mmol/l); can be excreted only by the kidneys. It filters freely in the glomeruli and is partially reabsorbed by passive mechanisms (see p. 255) about $^1/_3$ in the proximal tubule, and another $^1/_3$ in the medullary collecting duct, the latter stimulated by ADH, hence increasing the medullary hyperosmolality. In antidiuresis, urea reabsorption increases and C_{urea} falls.

Uric acid: Accounts for only 5% of the nitrogen eliminated (plasma level 3.0–7.0 mg/dl or 180–420 μmol/l); it is filtered in the glomeruli and secreted in the proximal tubule (see figure on p. 254), but there is also passive reabsorption in the proximal tubule and in the descending limb of the loop of Henle. The urine contains only ca. 10% of the filtered load. This relatively low rate of excretion normally prevents uric acid precipitation in the distal nephron or lower urinary tree.

Creatinine: Produced by muscle metabolism (plasma level 0.6–1.5 mg/dl or 60–130 μmol/l, only slight individual variations); is ultrafiltered; in normal man, a small amount is secreted in the proximal tubule and none is reabsorbed. Its clearance is practically the same as the inulin clearance, so that it can be used to estimate the GFR when the latter is normal. With a very low GFR, creatinine clearance overestimates the GFR.

Transport of protons (H⁺) and bicarbonate by the renal tubule
(see also acid–base status of the blood, beginning p. 207)

H⁺ ions ('fixed' or non-volatile): Produced by metabolism, ca. 60–100 mmol/d; can be excreted only via the kidneys. Three moieties:
- Excretion as free H⁺ (<1% of total amount) in the collecting duct (pH 4.5).
- Excretion as titratable acid (30–40 mmol/d), i.e. 'neutralized' by urine buffers; especially as phosphates or ketoacids produced in excess by metabolism.
- Excretion as NH_4^+ (ammonium); 30–60 mmol/d; can be increased 10-fold in metabolic acidosis. Here and in H⁺ excretion as titratable acid, there is *de novo* generation of one HCO_3^- per H⁺ excreted in the renal tubule. H⁺ secretion is via Na^+/H^+ exchange in proximal tubule and via H⁺ pump (H⁺ or H⁺, K⁺-ATPase) in collecting duct.

Bicarbonate: Filtered load 4500 mmol/d, reabsorption 100%, see p. 255. Fractional reabsorption amounting to ca. 60% of the load in the proximal tubule, capacity-limited reabsorption in distal segments (i.e. when too much bicarbonate is present, the excess is excreted).

Excretion of K⁺

For filtered load and excretion see p. 255; plasma level 3.4–5.2 mmol/l. Potassium is acquired in excess from food, especially with a meat-rich diet; excretion: 10% by the colon, 90% by the kidney; in the proximal tubule fractional reabsorption of 70–80% of the load, an additional 10% is reabsorbed in the loop of Henle. fine adjustment of excretion is under aldosterone control in the distal tubule and collecting duct, where K⁺ is usually secreted, but can be reabsorbed depending on homeostatic needs. The control of K⁺ balance by the kidney is related to and in many ways secondary to the control of Na⁺ balance. K⁺ excretion also depends on acid–base status: loss in alkalosis, retention in acidosis.

Cellular transport mechanisms in the proximal tubule and collecting duct

Transport of substances in the proximal tubule (see also p. 255)

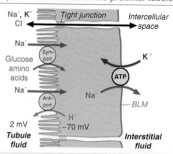

The basolateral membrane (BLM) contains the Na⁺, K⁺-ATPase. This pump maintains a low Na⁺ concentration in the tubule cell (primary active transport). The permeability of the BLM is low for Na⁺ and high for K⁺, so that the membrane voltage is cell-interior negative, typically –70 mV. Na⁺ can therefore enter the cell by an energetically downhill process, mostly via an Na⁺/H⁺ exchanger and via Na⁺/glucose and Na⁺/amino acid co-transporters.

The Na⁺ electrochemical gradient provides the driving force for active influx of amino acids and glucose (secondary active co-transport or symport) and H⁺ secretion (secondary active, electroneutral countertransport or antiport). Cl⁻ transport is partly transcellular (at the luminal membrane, by Cl⁻/organic anion [formate] exchange) and partly intercellular, by diffusion and solvent drag. Water reabsorption is osmotic, driven by the fall in luminal fluid osmolality and the increase in peritubular fluid osmolality because of solute reabsorption.

Ion transport in the thick ascending segment of the loop of Henle and in the collecting duct (see also p. 256)

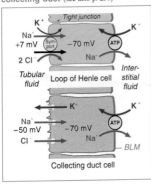

Active Na^+ transport is by the basolateral Na^+, K^+-ATPase (see p. 258)

In the loop of Henle cell, Na^+ and Cl^- enter via an electroneutral Na^+–K^+–$2Cl^-$ symport. Cl^- leaves the cells via Cl^- channels (and/or a KCl co-transporter). The luminal membrane is impermeable to water (see p. 255) so that NaCl reabsorption dilutes the luminal fluid and concentrates the peritubular fluid.

In the collecting duct Na^+ entry is by electrodiffusion across luminal membrane Na^+ channels, regulated by aldosterone (which increases the incorporation and/or activity of the channels). K^+ secretion occurs through apical membrane K^+ channels. The mechanisms of Cl^- transport are not fully understood.

Important cross-references

Micturition and bladder continence p. 106

Kidney function and circulation
    Role of the kidney in long-term regulation of blood pressure p. 189
    Renin-angiotensin-aldosterone mechanism pp. 143, 189

Acid–base metabolism
    Detailed description beginning p. 204

Water and electrolyte metabolism is described on p. 260

Hormones
    ADH p. 138
    Aldosterone p. 143
    Ca^{2+} and phosphate balance p. 264

Water balance

Distribution of total body water in adults (modified from Deetjen, 1989b)

Transcellular	1 liter	2 %
plasma	3 liters	7 %
Interstitium	13 liters	31 %
Cells	25 liters	60 %
Total	42 l	100 %

Transcellular: water in cerebrospinal fluid, chambers of the eye, GIT, exocrine glands, renal tubules and urinary tracts; plasma: water in the blood plasma (see also p. 147); interstitial space: extracellular water (including lymph, not including plasma); cells: intracellular water.

The volume of the fluid compartments is measured from the volumes of distribution of inert substances that distribute in specific compartments, including plasma. The concentration (e.g. in plasma) is measured and the volume (V) found from the concentration (C) and the infused quantity (Q) (corrected for excretion) by the relation V = Q/C:

1. Total body water: e.g. with heavy or tritium-labeled water (D_2O or THO)
2. Extracellular space: inulin or thiosulfate
3. Plasma volume: radioactive iodine, ^{131}I or Evans blue (both bind to plasma proteins)
4. Cell volume: space 1 minus space 2
5. Interstitial space: space 2 minus space 3.

Water content of individual organs (Skeleton, 1972) and relation between fat and water content of the body (Behnke, 1941/2)

Organ	% of body weight (BW)	Water content (%)	Water (liters) for 70 kg BW
Blood	8.0	83.0	4.65
Kidneys	0.4	82.7	0.25
Heart	0.5	79.2	0.28
Lungs	0.7	79.0	0.39
Spleen	0.2	75.8	0.10
Muscle	41.7	75.6	22.10
Brain	2.0	74.8	1.05
GIT	1.8	74.5	0.94
Skin	18.8	72.0	9.07
Bone	15.9	22.0	2.45
Fat	10–50	10.0	0.70

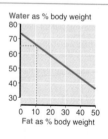

Water as % body weight

Fat as % body weight

The values above, obtained by indicator-dilution methods, are averages for the whole body. However, the water content of individual organs varies widely. Fat contains 10% water, all other organs (without fat) average 73%. Therefore from a measurement of total body water the proportion of fat in the body can be found by:

$$\% \text{ body fat} = 100 - \left[\% \text{ body water} / 0.73\right]$$

A young adult with 65% total water content thus has about 10% fat; in an obese person with 37% water content, fat accounts for 50% of the body weight.

Daily water turnover (fluid balance)

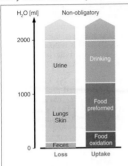

Daily water losses:
Urine (moderate antidiuresis)	1000 ml
Feces (minimum)	100 ml
Insensible losses	900 ml
Total	2000 ml

Daily water gains
Drinking	800 ml
Present in food	900 ml
Water of oxidation	300 ml
Total	2000 ml

Water turnover can vary considerably, depending on ambient and other conditions, and is particularly likely to increase (see p. 225). The minimum urine volume in normal people is about 500 ml/d. The maximum is over 10 l/d.

The substances that must be excreted in the urine together amount to about 600 mosmol/d; because the kidneys can concentrate urine to about 1300 mosmol/kg (see p. 256), 0.5 liter of water is required to dissolve these substances. Insensible water losses occur by evaporation from the skin and airways (see p. 221) and do not include sweat. A mixed diet consists on average of 65% water as such and oxidation of the food produces water in varying amounts: 1 ml water per gram of fat, 0.6 ml per gram of carbohydrate and 0.4 ml per gram of protein; together, on average, these provide the above 300 ml/d.

Reabsorption of water from the GIT

Ingestion of pure water does not cause rapid absorption because it is made almost isosmotic in the stomach (secretion of HCl) and duodenum (secretion of Na^+ and HCO_3^-) and then, mainly in the jejunum, is slowly reabsorbed secondarily to active absorption of NaCl (mechanism like that in the proximal tubule, see pp. 255, 258).

For the survey of the regulation of water balance on the next page, the following cross-references on other aspects of water and electrolyte balance are important:

- **Thirst** (p. 133): the adequate stimuli for the sensation of thirst, the intra- and extracellular sensors and the neuronal and hormonal systems involved are the same as are discussed here.

- **Posterior pituitary** (p. 138): certain aspects of ADH (antidiuretic hormone, vasopressin) functions are discussed there.

- **Long-term regulation of the circulation** (p. 189): survey of the mechanisms that regulate extracellular volume and hence the degree of filling of the vascular system.

- **Thermoregulation** (p. 219): in particular, heat loss by evaporation of sweat, heat acclimation in a tropical climate.

Regulation of water balance

Process	Description/comments
Osmoregulation in the liver	The water absorbed from the GIT may be moderately hyposmotic (see p. 261). Some of it reaches the liver via the portal vein (physiological liver swelling, up to 30%). This reduces the tonic discharge of the hepatic osmosensors, which reflexly diminishes ADH synthesis and release, initiating water diuresis (see p. 138)
Osmoregulation in the hypothalamus	If the blood becomes hyposmotic, there is: (a) inhibition of osmosensors in the hypothalamus, with reduced production and release of ADH and water diuresis (see above), (b) cessation of the sensation of thirst. Plasma hyperosmolality (water deficit) has the opposite effects: (a) increased ADH synthesis and secretion, with antidiuresis, and (b) thirst
Volume regulation	Large reductions in extracellular fluid volume are detected by the volume sensors in the low-pressure system, and also by presso-sensors (see p. 188) causing both ADH secretion (antidiuresis) and thirst (see p. 133)

Electrolyte balance

Electrolyte concentrations in plasma and intracellular fluid (plasma values include normal range) (modified from Deetjen, 1989b)

Electrolyte (ion)	Plasma (mmol/l)	Cell (mmol/l)
Na^+	142 (130–155)	10
K^+	4 (3.2–5.5)	155
Ca^{2+}	2.5 (2.1–2.9)	<0.001[a]
Mg^{2+}	0.9 (0.7–1.5)	15
Cl^-	102 (96–110)	8
HCO_3^-	25 (23–28)	10
HPO_4^{2-}	1 (0.7–1.6)	65[b]
SO_4^{2-}	0.5 (0.3–0.9)	10
Proteins	2	6
Organic acids	4	2

[a] Free Ca^{2+} in cytosol.
[b] Including organic phosphates.

In spite of the differences in ion distribution, the osmotic pressures of the intra- and extracellular fluid are identical under steady-state conditions.
The main cause of ionic disequilibrium between intra- and extracellular fluids is the activity of the Na^+, K^+-ATPase (see p. 11), which keeps the concentration of Na^+ inside the cell low and the concentration of K^+ high (the K^+ gradient is the main determinant of resting membrane potentials; see p. 11).

Sodium and potassium balances

Sodium: Total content for 70 kg body weight: 4200 mmol, of which $1/_3$ is fixed as part of the crystalline bone structure; the rest is exchangeable with the sodium in the extracellular space. Only ca. 2.5% is in the cells of soft tissues. Intake in food: 160 mmol/d, ca. 5% of the exchangeable amount (for renal load and excretion see p. 255).
Potassium: Total content 3300 mmol, almost all of which is exchangeable. Only ca. 2.5% is in the extracellular space. In general, with a present-day mixed diet, potassium is consumed in excess (for load and excretion see p. 255, excretory mechanism p. 258).

Regulation of salt balance (extracellular fluid volume)

Process	Description/comments
First factor	Depletion of extracellular fluid produces a reduction in blood volume on the arterial side ('effective blood volume', EBV), renal blood flow and GFR fall and renal losses of salt and water decrease
Second factor	The same stimulus stimulates renin secretion by the juxtaglomerular apparatus; renin hydrolyzes angiotensinogen, to produce angiotensin I, which by the action of converting enzyme becomes angiotensin II; angiotensin II stimulates secretion of aldosterone; aldosterone causes salt retention by the kidney, and indirectly water retention as well
Third factor(s)	Several other mechanisms tend to cause salt and water retention following hypovolemia. The most prominent are ANF (less distension of atria $\rightarrow$ less secretion of ANF $\rightarrow$ less Na^+ loss), and catecholamines (secretion stimulates reabsorption in the proximal tubule).

The situations described above correspond to physiological responses to a fall in extracellular fluid volume; the opposite mechanisms operate in case of an increase in fluid volume provided that there are changes in EBV in the same direction. In some pathophysiological conditions this does not occur, e.g. in congestive heart failure the extracellular fluid volume is generally elevated, but the EBV is low, because of low cardiac output. Hence the kidneys inappropriately retain salt and water.

Factors determining K^+ distribution and renal K^+ excretion

Process	Factors
K^+ distribution between body fluid compartment	High intracellular $[K^+]$ is maintained by Na^+, K^+-ATPase. Factors tending to elevate $[K^+]$ are catecholamines, insulin, high intracellular $[K^+]$, alkalosis and plasma hyperosmolality. Opposite changes in these factors lower intracellular $[K^+]$.
Regulation of K^+ secretion	Occurs by modulation of secretion by the cortical collecting tubule. Factors increasing K^+ secretion are: high plasma $[K^+]$, aldosterone, vasopressin, high flow rate in the distal nephron, high luminal $[Na^+]$, low luminal $[Cl^-]$ and alkalosis. Opposite changes in these factors tend to reduce K^+ secretion.

Calcium homeostasis (modified from Ziegler and Minne, 1991)

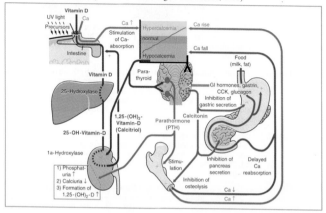

Three hormones act in co-ordination to regulate calcium balance; when calcium intake is excessive, the mechanisms below turn off

Hormone	Action/comments
PTH	A reduced Ca level stimulates the secretion of parathyroid hormone (PTH) from the parathyroid glands, eliciting mobilization of Ca and phosphate from bone by activation of osteoclasts. At the same time, Ca excretion by the kidneys is decreased, and phosphate excretion is increased; the formation of active vitamin D in the kidneys is also stimulated (see below)
Calcitonin	Is released from the C-cells in the thyroid gland when the plasma free Ca level is elevated. Osteoclasts are inhibited, more Ca and phosphate are incorporated into bone, and absorption of Ca from the intestine is reduced. Effective only in acute, not in chronic changes of blood Ca level ('acute hormone')
Active vitamin D	Active vitamin D, 1,25-dihydroxycholecalciferol; produced in the kidneys from vitamin D. Production is stimulated by PTH (see above); active vitamin D potentiates the actions of PTH by increasing intestinal Ca absorption (by induction of the Ca-binding protein)

The organism needs all three hormone systems only in negative Ca balance or with low plasma free Ca; if these parameters are normal, the hormonal systems are inactive. They act synergistically and are briefly antagonistic only in individual systems (e.g. at osteoclasts). The regulation of phosphate balance is much less precise than that of calcium.

Synthetic pathways of the gonadal steroid hormones (modified from Wuttke, 1989b)

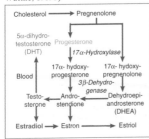

The sex hormones are steroids (see p. 137); they are all synthesized by a common pathway and are present in both males and females, but some of them in very different amounts:

Corpus luteum and placenta produce chiefly the gestagen progesterone.

Ovaries produce estrogens (mainly estradiol).

Leydig cells produce androgens (mainly testosterone); the latter is sometimes not converted to the active 5α-dihydrotestosterone (DHT) until it reaches the target organ (e.g. prostate, seminal vesicle).

Hormonal regulation of male sexual functions

Hypothalamic–pituitary–testicular control circuit (modified from Wuttke, 1989b)

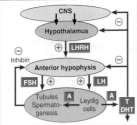

In the hypothalamus LHRH (syn.: GnRH) is synthesized (see p. 139) and released in pulses every 2–4h; in the anterior pituitary it elicits release of both LH and FSH (see p. 139).

• LH stimulates the Leydig cell(s) to increase production of androgens (A), esp. testosterone (T); these provide negative feedback to the anterior, not posterior pituitary and hypothalamus (closing the control loop; DHT [see above] has no feedback action); the A, esp. T, act on spermatogenesis and on many target organs (see below).

• FSH also acts on spermatogenesis in the convoluted seminiferous tubules (see below), and simultaneously induces the Sertoli cells to produce the peptide hormone inhibin, which has a negative feedback effect on FSH secretion.

Sites of synthesis and transport of the androgens, esp. testosterone; actions on target organs other than genital organs

• Synthesis in Leydig interstitial cells (see above), in zona reticularis of the adrenal cortex (see p. 142) and in the ovary in women; carrier protein in the blood (see p. 137) is the sex hormone-binding globulin (SHBG).

• Main extragenital biochemical action is to stimulate protein synthesis; i.e. they are anabolic; this includes promoting bone formation by stimulation of the protein matrix, also increase in muscle mass, especially under training (anabolic steroids are androgen derivatives).

• Influence on sexual behavior slight in adult and elderly men, as long as a minimal amount (350–1000 ng/l blood) is present.

Spermatogenesis (duration ca. 70 d) in the convoluted seminiferous tubules of the testis and maturation of the sperm cells in the epididymis is under the control of FSH and testosterone and proceeds as follows:

- The germ cells in the tubule wall continually generate spermatogonia, which divide by mitosis into an inactive reserve cell (for a later division into active and inactive cells) and an active spermatogonium.

- Active spermatogonium divides by mitosis into two primary spermatocytes, each of which again divides (first meiotic division) into two secondary spermatocytes.

- Secondary spermatocytes in turn undergo the second meiotic division, to form round spermatids, each of which contains 23 chromosomes.

- Spermatids mature without further division, becoming the elongated spermatozoa (or simply sperm); this process is called spermiogenesis.

- The non-motile sperm are washed into the ca. 5 m-long epididymal ducts, where they spend 5–12 d in final maturation, in particular acquiring motility.

- The sperm are stored (in some cases for months) in small numbers in the epididymis but mostly in the vas deferens and the ampulla; the temperature of the scrotal organs, 2°C below the core body temperature, is necessary for optimal spermatogenesis.

The sperm are most motile in a slightly alkaline milieu, and the seminal fluid (ejaculate) is composed accordingly; its components originate in the vas deferens, the seminal vesicle and the prostate; not until ejaculation is the ejaculate (2–6 ml, 40–250 million sperm/ml, pH 7.3–7.8) assembled from these sources and the urethral and bulbourethral glands; hence it exists as this mixture only outside the body.

Hormonal regulation of female sexual functions

Hypothalamic–pituitary–ovarian control system (modified from Wuttke, 1989b)

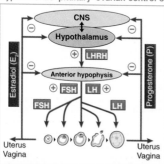

- LHRH, as in the man (see above), is synthesized in the hypothalamus and released in pulses; the difference from male LHRH release is that in the man the pulsation is uniformly distributed in time, whereas in the woman it has a monthly rhythm: in the first half of the menstrual cycle the pulses occur ca. every 90 min, and after ovulation every 2–4 h (as normally in men).

- In the AP LHRH causes the release of FSH and LH, which act on the ovary.

The feedback mechanisms controlling secretion of the hormones produced in the ovary, especially estradiol (E_2) and progesterone, to the LHRH neurons of the hypothalamus and the FSH and LH cells of the AP differ from those in the control systems considered previously; because of the rhythmic production of the hormones involved, feedback is not constant but fluctuates in a 28 d rhythm, discussed on the next page.

The pulsatile pattern of LH release is observed only in adult women with functional ovaries. The pulse generator is located in the medial–basal hypothalamus. Estrogens supress LH pulses (mediated by NE) and also inhibit the AP. Progesterone also influences pulsatile LH release (action on the brain, mediated by opioid peptides).

Hormonal, ovarian and uterine changes in the course of a menstrual cycle (modified from Wuttke, 1989b)

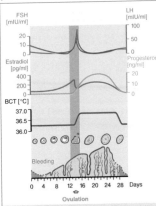

Subdivisions of the cycle

1. Follicular phase (days 1–13, counting from the first day of menstrual period)
2. Ovulatory phase (days 13–14)
3. Luteal phase (days 14–28)

1: During the menstrual period, estrogen and progesterone levels are low (corpus luteum regression, follicle immaturity). Following the onset of menstrual period FSH and LH (pulses of low amplitude and progressively increasing frequency) stimulate the maturation of numerous follicles; toward the middle of the follicular phase one of them becomes dominant, suppresses all others and under the influence of FSH produces increasing amounts of estradiol E_2, which (i) induces proliferation of the endometrium (hyperplasia and hypertrophy), (ii) sensitizes the FSH and LH cells of the pituitary to LHRH and (iii) inhibits the release of FSH.

2: In the middle of the cycle, when sensitization of pituitary cells has occurred, an increased release of LHRH stimulates LH and FSH release; now, under influence of LH (LH surge), the following occur: (i) ovulation, (ii) the conversion of the follicle to corpus luteum and (iii) production and secretion of progesterone (P) (role of FSH at this time is unclear). LH surge requires estrogen (in a narrow concentration range). LHRH has a permissive role.

3: E_2 and esp. P exert negative feedback on the AP, reducing the production of LH (LH pulse frequency is reduced) and FSH; P converts the proliferating endometrium to a secretory organ (mucus secretion) and raises the basal body temperature (BBT) by ca. 0.5°C; the corpus luteum has a predetermined life-span of 14 d; the P production falls, the endometrium becomes ischemic, undergoes regression and menstrual bleeding begins (i.e. it is P withdrawal bleeding); the cycle has been completed.

Additional aspects of the menstrual cycle; conception

- **Two-cell theory of ovarian steroid production:** In the follicular phase the thecal cells (originate in the ovarian stroma), under the influence of LH, synthesize androgens, esp. testosterone, which are converted to estrogens, esp. estradiol, in the granulosa cells under the influence of FSH; after luteinization the granulosa cells, under the influence of LH, synthesize large amounts of progesterone.
- **Female fertility:** After ovulation the egg is fertilizable for ca. 24 h; fertilization generally occurs in the Fallopian tubes; hence intercourse as early as 48 h before ovulation can result in fertilization, but fertilization is extremely unlikely for intercourse >24 h after ovulation
- **Cervical mucus:** Its secretion changes during the menstrual cycle: during follicular phase mucus is abundant, alkaline, viscous and very elastic, and forms a fern pattern when placed on a glass slide; elasticity peaks at ovulation time; during luteal phase mucus secretion decreases, elasticity decreases and no fern pattern is formed.
- **Oogenesis:** The prophase of the 1st maturation division occurs in the fetal period and the process is then interrupted for 12–50 years; after puberty (see below) a cohort of egg cells develop further in each follicular phase (see above) and in the egg of the dominant follicle the second maturation division continues into metaphase; only if this egg is fertilized does the second division proceed to completion.

Male and female puberty, menopause

Puberty is the maturation of reproductive ability; it coincides with the onset of pulsatile LHRH release; begins in males at age 11–13; associated with:

- Distinct rise in FSH and LH secretion; FSH stimulates spermatogenesis (see p. 266) and production of inhibin (by Sertoli cells), which inhibits FSH secretion. LH stimulates the production of androgens (for sites of synthesis, carrier proteins and extragenital actions see p. 265); testosterone exerts negative feedback by slowing LHRH pulses.
- Physical and psychological masculinization results mainly from the elevated androgen level; includes among other things: a spurt of linear growth followed by ossification of the epiphyseal bone regions, lowering of the voice by growth of the larynx, and development of masculine hair pattern (due to DHT, see p. 265); growth hormone contributes to linear growth.

In females puberty begins between ages 9 and 11 years; associated with:

- Maturation of follicles due to increasing release of FSH and LH as a result of the beginning of pulsatile LHRH release; these follicles produce estrogens, especially estradiol. With FSH and estradiol acting together, there is expression of LH receptors and secretion of progesterone and prostaglandin starts. The capacity of FSH to stimulate estrogen production diminishes, hence there is an abrupt decrease in estrogen, which precedes ovulation.
- Anovulatory cycles in the early phase of puberty (reason unknown); therefore the early menstrual bleeding is purely estrogen withdrawal bleeding; regular cycles begin in the late stage of puberty.
- Development of secondary sexual characteristics with physical and psychological feminization especially under the influence of estrogens (are distinctly less anabolic than the androgens); includes breast development, puberal growth spurt, widening of the pelvis. Progesterone raises body temperature and may elicit mood changes; withdrawal of progesterone triggers menstruation.

Menopause is defined as the time when the last normal menstrual cycle occurs (from about age 45 to 55)

The main cause is the decrease in the number of follicles available for hormonal stimulation; hence ovarian estrogen production ceases; the endocrine functions of hypothalamus and anterior pituitary are retained but there is no feedback, and hence the FSH and LH levels in the blood rise sharply; because of the lack of estrogens, there is loss of vaginal epithelium, decreased breast size, and loss of bone matrix, causing osteoporosis.

Pregnancy, birth, lactation

Factors involved in the occurrence of pregnancy (modified after Rabe, 1990)

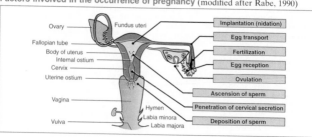

Hormonal aspects of pregnancy (duration 280 d, 40 weeks)

- **Preservation of the endometrium:** The trophoblast (outer wall of the blastocyst) produces human chorionic gonadotropin (hCG), which acts like LH and stimulates the corpus luteum to synthesize and secrete progesterone; therefore menstrual flow is suppressed (often the first sign of a pregnancy).

- **Nidation and placentation:** On arrival in the uterus (6–8 d after fertilization in a tube) the trophoblast produces proteolytic enzymes that allow it to become implanted in the endometrium (nidation); at the same time there is a further increase in production of hCG and hence of progesterone; after the placenta has fully developed it takes over the production of progesterone (from about the 8th or 19th week on).

- **Hormone production by the fetus:** From the third month on, the fetal adrenal medulla produces increasing amounts of DHEA, which is converted to estriol in the placenta; estriol is excreted in the mother's urine and serves an indicator (normal limits) that the pregnancy is proceeding normally; human placental lactogen (hPL) is also formed, but its role is unclear (anti-insular action, lipolysis, no effect on milk production).

- **Preparation for lactation:** Fetal estrogen stimulates the secretion of prolactin from the AP of the mother; this prepares the mammary glands for lactation (see below); prior to birth estrogen antagonizes the effect of prolactin on the mammary glands.

Hormonal control of delivery and nursing

- **Initiation of labor:** Oxytocin action on the sensitized uterus (Ferguson's reflex, see p. 138); prostaglandins relax the cervix.

- **Delivery:** Also by Ferguson's reflex, i.e. uterine contraction, stimulated by increased release of oxytocin.

- **Hormonal control of lactation:** see p. 140 (actions and regulation of prolactin) and p. 138 (milk ejection reflex).

References

Altner, H. (1985a) Physiologie des Geschmacks, in *Grundriß der Sinnesphysiologie*, 5th edn (ed. R.F. Schmidt), Springer, Heidelberg.

Altner, H. (1985b) Physiologie des Geruchs, in *Grundriß der Sinnesphysiologie*, 5th edn (ed. R.F. Schmidt), Springer, Heidelberg.

Altner, H. and Boeckh, J. (1990) Geschmack und Geruch, in *Physiologie des Menschen*, 24th edn (eds R.F. Schmidt and G. Thews), Springer, Heidelberg.

Antoni, H. (1989) In *Human Physiology*, 2nd edn (eds R.F. Schmidt and G. Thews), Springer, Heidelberg.

Aschoff, J. and Wever, R. (1958) Kern und Schale im Wärmehaushalt des Menschen. *Naturwissenschaften*, **45**, 477.

Asmussen, E. and Nielsen, M. (1955) *Physiol. Rev.*, **35**, 778.

Behnke, A.R. (1941/2) Physiological studies pertaining to deep sea diving and aviation, especially in relation to fat content and composition of the body. *Harvey Lectures*, **37**, 198.

Bevegard, S. *et al.* (1960) *Acta Physiol. Scand.*, **49**, 279.

Birbaumer, N. and Schmidt, R.F. (1996) *Biologische Psychologie*, 3rd edn, Springer, Heidelberg.

Boothby, W.M., Berkson, J. and Dunn, H.L. (1936) *Am. J. Physiol.*, **116**, 468.

Broemser, P. (1938) *Kurzgefasstes Lehrbuch der Physiologie*, Thieme, Leipzig.

Bro-Rasmussen *et al.* (1956) *Acta Physiol. Scand.*, **37**, 185.

Brück, K. (1989) Thermal balance and the regulation of body temperature, in *Human Physiology*, 2nd edn (eds R.F. Schmidt and G. Thews), Springer, Heidelberg.

Darian-Smith *et al.* (1973) *J. Neurophysiol.*, **36**, 325.

Deetjen, P. (1989a) The function of the kidneys, in *Human Physiology*, 2nd edn (eds R.F. Schmidt and G. Thews), Springer, Heidelberg.

Deetjen, P. (1989b) Water and electrolyte balance, in *Human Physiology*, 2nd edn (eds R.F. Schmidt and G. Thews), Springer, Heidelberg.

Deetjen, P. and Kramer, K. (1961) *Pflügers Arch.*, **273**, 636.

de Groat, W.C. (1975) *Brain Res.*, **87**, 201–213.

Dubois, E.F. (1937) *The Mechanism of Heat Loss and Temperature Regulation*, Stanford University Press, Stanford, CA.

Dudel, J. (1985) Allgemeine Sinnesphysiologie, Psychophysik, in *Grundriß der Sinnesphysiologie*, 5th edn (ed. R.F. Schmidt), Springer, Heidelberg.

Dudel, J. (1990) Erregungsübertragung von Zelle zu Zelle, in *Physiologie des Menschen*, 24th edn (eds R.F. Schmidt and G. Thews), Springer, Heidelberg.

Dudel, J. (1995) Informationsvermittlung durch elektrische Erregung, in *Physiologie des Menschen*, 26th edn (eds R.F. Schmidt and G. Thews), Springer, Heidelberg.

Eccles, J.C. (1964) *The Physiology of Synapses*, Springer, Heidelberg.

Eccles, J.C. (1969) *The Inhibitory Pathways of the Central Nervous System*. The Sherrington Lectures IX, Thomas, Springfield, IL.

Ervin, F.R. and Anders, T.R. (1970) Normal and pathological memory: data and a conceptual scheme, in *The Neurosciences: Second Study Program* (ed. F.O. Schmitt), Rockefeller University Press, New York.

Ewe, K. and Karbach, U. (1989) In *Human Physiology*, 2nd edn (eds R.F. Schmidt and G. Thews), Springer, Heidelberg.

Furness, J.B. *et al.* (1992) Roles of peptides in transmission in the enteric nervous system, *Trends Neurosci.*, **15**, 66.

Gauer, O.H. (1972) Kreislauf des Blutes, in *Physiologie des Menschen. Bd. 3: Herz und Kreislauf* (eds O.H. Gauer, K. Kramer and R. Jung), Urban & Schwarzenberg, München.

Golenhofen, K. (1981) *GK1 Physiologie*, 9th edn, VCH, Weinheim.

Gray, G. (1982) *Neuropsychology of Anxiety*, Oxford University Press, Oxford.

Grote, J. (1989) Tissue respiration, in *Human Physiology*, 2nd edn (eds R.F. Schmidt and G. Thews), Springer, Heidelberg.

Grüsser, O.-J. and Grüsser-Cornehls, U. (1990) In *Physiologie des Menschen*, 24th edn (eds R.F. Schmidt and G. Thews), Springer, Heidelberg.

Guyton, A.C. (1976) *Textbook of Medical Physiology*, 5th edn, Saunders, Philadelphia.

Haden, R.L. (1935) *Am. J. Clin. Pathol.*, **5**, 354.

Hassler, R. (1979) Die neuronalen Steuerungssysteme für Wachsein, Schlaf und Bewußtseinsvorgänge, in *Der bewußtlose Patient* (eds F.W. Ahnefeld *et al.*), Springer, Heidelberg.

Hatt, H. (1993) Physiologie des Geruchs, in *Neuro- und Sinnesphysiologie* (ed. R.F. Schmidt), Springer, Heidelberg.

Henry, J.P. and Gauer, O.H. (1950) *J. Clin. Invest.*, **29**, 855.

Hobson, J.A. *et al.* (1986) Evolving concepts of sleep cycle generation: from brain centers to neuronal populations. *Behav. Brain Sci.*, **9**, 371–448.

Hodgkin, A.L. and Huxley, A.F. (1952) Quantitative description of membrane current and its application to conduction and excitation in nerve. *J. Physiol. (Lond.)*, **117**, 500.

Hutter, O.F. and Trautwein, W. (1956) *J. Gen. Physiol.*, **39**, 715.

Hurtado, A. (1964) Animals in high altitudes: resident man, in *Handbook of Physiology. Section 4: Adaptation on the Environment* (ed. D.B. Dill), American Physiological Society, Washington, DC.

Jänig, W. (1987) Vegetatives Nervensystem, in *Grundriß der Neurophysiologie*, 6th edn (ed. R.F. Schmidt), Springer, Heidelberg.

Jänig, W. (1995) Vegetatives Nervensystem, in *Physiologie des Menschen*, 26th edn (eds. R.F. Schmidt and G. Thews), Springer, Heidelberg.

Jovanović, U.J. (1971) *Normal Sleep in Man*, Hippokrates, Stuttgart.

Kandel, E.R. *et al.* (eds) (1991) *Principles of Neural Science*, 3rd edn, Elsevier, Amsterdam.

Kenshalo, D.R. (1976) In *Sensory Functions of the Skin in Primates* (ed. Y. Zotterman), Pergamon, Oxford.

Keul *et al.* (1965) *Pflügers Arch. Ges. Physiol.*, **282**, 1.

Keul , J., Doll, E. and Keppler, D. (1969) *Muskelstoffwechsel*, Barth, München.

Klinke, R. (1990) Gleichgewichtssinn, Hören, Sprechen, in *Physiologie des Menschen*, 24th edn (eds R.F. Schmidt and G. Thews), Springer, Heidelberg.

Kolb, B. and Winshaw, I.Q. (1985) *Fundamentals of Human Neuropsychology*, Freeman, New York.

Landis, E.M. and Pappenheimer, J.R. (1963) *Handbook of Physiology. Section II: Circulation II*, American Physiological Society, Washington, DC.

Lundberg, J.M. (1981) *Acta Physiol. Scand.*, **112** (Suppl. 496), 1–57.

Mayer, C.J. (1974) *Entladungsmuster und Funktion von Ganglienzellen des Auerbachschen Plexus*, Habilitationsschrift, München.

Milnor, W.R. (1974) In *Medical Physiology*, 13th edn (ed. V.B. Mountcastle), Mosby, St Louis.

Müller, S. (1991) *Memorix Spezial Notfallmedizin*, VCH, Weinheim.

National High Blood Pressure Education Program Coordinating Committee (1990) *Arch. Intern. Med.*, **150**, 2270.

Nicoll, R.A., Kauer, J.A. and Malenka. R.C. (1988) The current excitement in long-term potentiation. *Neuron*, **1**, 97–103.

Patton, H.D. *et al.* (1989) *Textbook of Physiology*, 21st edn, Saunders, Philadelphia.

Peachey, L.D. *et al.* (eds) (1983) *Handbook of Physiology. Section 10: Skeletal Muscle*, American Physiological Society, Bethesda, MD.

Penfield, W. and Rasmussen, T. (1950) *The Cerebral Cortex of Man*, Macmillan, New York.

Picton *et al.* (1974) *J. Electroenceph. Clin. Neurophysiol.*, **36**, 179.

Piiper, J. (1975) In *Physiologie des Menschens*, Vol. 6, 2nd edn (eds O.H. Gauer, K. Kramer and R. Jung), Urban & Schwarzenberg, München.

Rabe, T. (1990) *Gynäkologie und Geburtshilfe*, VCH, Weinheim.

Reichlin, S., Baldessarini, R.J. and Martin, J.B. (1978) *The Hypothalamus*, Raven, New York.

Rein, H. and Schneider, M. (1971) *Einführung in die Physiologie des Menschen*, 16th edn, Springer, Heidelberg.

Reindell, H., König, K. and Roskamm, H. (1967) *Funktionsdiagnostik des gesunden und kranken Herzens*, Thieme, Stuttgart.

Richter, P.W. (1986) *Zur Rhythmogenese der Atmung: Physiologie aktuell*, Vol. 1, Fischer, Stuttgart.

Roffwarg, H.P., Muzio, J.N. and Dement, W.C. (1966) Ontogenetic development of the human sleep-dream cycle. *Science*, **152**, 604.

Rohmert, W. (1962) *Untersuchung über Muskelmüdung und Arbeitsgestaltung*, Beuth, Berlin.

Rüdel, R. (1993) Muskelphysiologie, in *Neuro- und Sinnesphysiologie* (ed. R.F. Schmidt), Springer, Heidelberg.

Ruff, S. and Strughold, H. (1957) *Grundriß der Luftfahrtmedizin*, Barth, München.

Schiller, L.R. (1983) Motor functions of the stomach, in *Gastrointestinal Disease* (eds M.H. Sleisenger and J.S. Fordtran), Saunders, Philadelphia.

Schmidt, R.F. (1983) *Medizinische Biologie des Menschen*, Piper, München.

Schmidt, R.F. (1990a) Nociception und Schmerz, in *Physiologie des Menschen*, 24th edn (eds R.F. Schmidt and G. Thews), Springer, Heidelberg.

Schmidt, R.F. (1990b) Integrative Leistungen des Zentralnervensystems, in *Physiologie des Menschen*, 24th edn (eds R.F. Schmidt and G. Thews), Springer, Heidelberg.

Schmidt, R.F. (1990c) Durst und Hunger: Allgemeinempfindungen, in *Physiologie des Menschen*, 24th edn (eds R.F. Schmidt and G. Thews), Springer, Heidelberg.

Schmidt, R.F. and Thews, G. (eds) (1989) *Human Physiology*, 2nd edn, Springer, Heidelberg.

Schmidt, R.F. and Wiesendanger, M. (1990) Motorische Systeme, in *Physiologie des Menschen*, 24th edn (eds R.F. Schmidt and G. Thews), Springer, Heidelberg.

Schulz, H., Pollmächer, T. and Zulley (1991) Schlaf und Traum, in *Pathophysiologie des Menschen* (eds K. Hierholzer and R.F. Schmidt), VCH, Weinheim.

Shepherd, G.M. (1983) *Neurobiology*, Oxford University Press, New York.

Siggard-Andersen, O. (1963) *Clin. Lab. Invest.*, **15**, 211.

Simeone and Lampson (1937) *Ann. Surg.*, **106**, 413.

Simon, E. and Meyer, W.W. (1958) *Klin. Wochenschr.*, **36**, 424.

Skeleton, H. (1972) The storage of water by various tissues of the body. *Arch. Intern. Med.*, **40**, 140.

Söling, H. (1991) Stoffwechselerkrankungen, in *Pathophysiologie des Menschen* (eds K. Hierholzer and R.F. Schmidt), VCH, Weinheim.

Sonnenblick, E.H. (1962) *Fed. Proc.*, **21**, 975.

Stellar, J.R. and Stellar, E. (1985) *The Neurobiology of Motivation and Reward*, Springer, Heidelberg.

Thaysen, J.H., Thorn, N.A. and Schwartz, I.L. (1954) Excretion of Na, K, Cl and HCO_3 in human parotid saliva. *Am. J. Physiol.*, **178**, 155.

Thews, G. (1989) Pulmonary respiration, in *Human Physiology*, 2nd edn (eds R.F. Schmidt and G. Thews), Springer, Heidelberg.

Thews, G. (1963) Der Transport der Atemgase. *Klin. Wochenschr.*, **41**, 120.

Thurau, K. and Wober, E. (1963) Zur Lokalisation der autoregulativen Widerstandsänderung in der Niere. *Pflügers Arch.*, **274**, 553.

Töndury, G. (1969) *Angewandte und topographische Anatomie*, 2nd edn, Thieme, Stuttgart.

Ulmer, H.V. (1989a) Energy balance, in *Human Physiology*, 2nd edn (eds R.F. Schmidt and G. Thews), Springer, Heidelberg.

Ulmer, H.V. (1989b) Work physiology, in *Human Physiology*, 2nd edn (eds R.F. Schmidt and G. Thews), Springer, Heidelberg.

Wade, O.L. and Bishop, M. (1962) *Cardiac Output and Regional Blood Flow*, Blackwell, Oxford.

Weiss, C. and Jelkmann, W. (1989) Functions of the blood, in *Human Physiology*, 2nd edn (eds R.F. Schmidt and G. Thews), Springer, Heidelberg.

Wenzel, H.G. and Piekarski, C. (1982) Klima und Arbeit: *Bayerisches Staatsministerium für Arbeit und Sozialordnung*, Eigenverlag, München.

Wiggers, C.J. (1952) *Circulatory Dynamics*, Lea & Febiger, Philadelphia.

Wiggers, C.J. (1954) *Physiology in Health and Disease*, Lea & Febiger, Philadelphia.

Willis, W.D. (1985) *The Pain System*, Karger, Basel.

Willis, W.D. and Coggeshall, R.E. (1991) *Sensory Mechanisms of the Spinal Cord*, 2nd edn, Plenum Press, New York.

Wintrobe, M.M. (1981) *Clinical Hematology*, 8th edn, Lea & Febiger, Philadelphia.

Witzleb, E. (1989) In *Human Physiology*, 2nd edn (eds R.F. Schmidt and G. Thews), Springer, Heidelberg.

Wuttke, K. (1989a) Endocrinology, in *Human Physiology*, 2nd edn (eds R.F. Schmidt and G. Thews), Springer, Heidelberg.

Wuttke, K. (1989b) Sexual functions, in *Human Physiology*, 2nd edn (eds R.F. Schmidt and G. Thews), Springer, Heidelberg.

Ziegler, R. and Minne, H.W. (1991) Hormone der Kalziumhomöostase, in *Pathophysiologie des Menschen* (eds K. Hierholzer and R.F. Schmidt), VCH, Weinheim.

Zimmermann, M. (1990) Das somatoviszerale sensorische System, in *Physiologie des Menschen*, 24th edn (eds R.F. Schmidt and G. Thews), Springer, Heidelberg.

Index